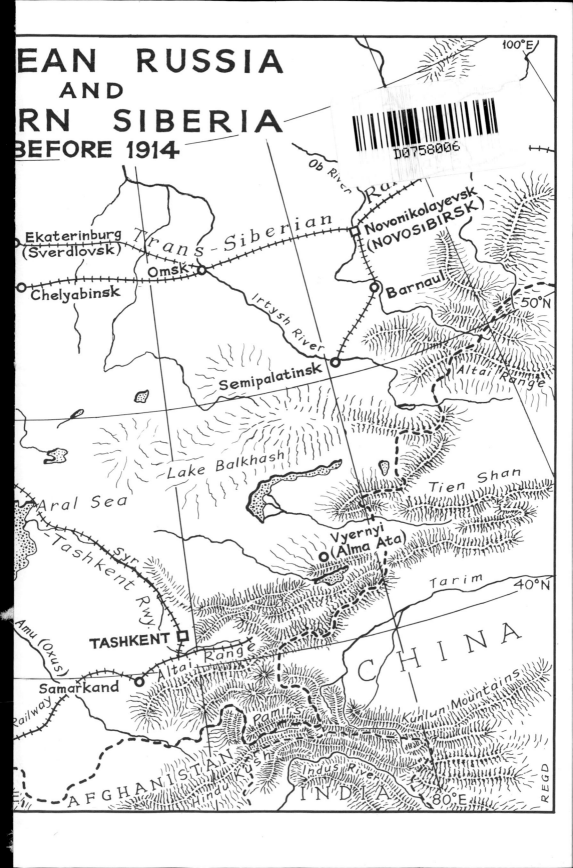

EAN RUSSIA
AND
RN SIBERIA
BEFORE 1914

100°E

Ob River

Trans-Siberian Ra...

Ekaterinburg
(Sverdlovsk)

Novonikolayevsk
(NOVOSIBIRSK)

Omsk

Barnaul

Chelyabinsk

50°N

Irtysh River

Altai Range

Semipalatinsk

Lake Balkhash

Aral Sea

Tien Shan

Syr

Vyernyi
(Alma Ata)

Tarim

40°N

Tashkent Rwy.

TASHKENT

CHINA

Amu (Oxus)

Altai Range

Kunlun Mountains

Samarkand

Railway

Pamirs

Indus River

80°E

AFGHANISTAN

Hindu Kush

INDIA

REGD

REMINISCENCES OF A RUSSIAN PILOT

Gatchina Days

Alexander Riaboff

Von Hardesty, Editor

SMITHSONIAN INSTITUTION PRESS
Washington, D.C.

Library of Congress Cataloging-in-Publication Data
Riaboff, Alexander, 1895–1984.
  Gatchina days.
    1. Riaboff, Alexander, 1895–1984. 2. Soviet Union—
History—Revolution, 1917–1921—Personal narratives.
3. Air pilots, Military—Soviet Union—Biography.
4. Aeronautics, Military—Soviet Union. I. Hardesty,
Von, 1939–   . II. Title.
DK265.7.R53A34   1986   947.084′092′4   85-26247
ISBN 0-87474-802-X

The maps in this book were drawn by R.E.G. Davies,
curator, National Air and Space Museum.

All photographs are from the Riaboff Collection except for
those on the following pages from the files of the National
Air and Space Museum: 10, 17, 18, 20, 21, 27, 31, 50
(top), 88, 94, 98, 101, 102, 111.

COVER: Alexander Riaboff at the controls of a
Nieuport 17 fighter.

FRONTISPIECE: Alexander Riaboff (right) poses in
a repair hangar with his flight instructor (center)
and Valerian Przhegodsky (left), another cadet
and a close friend.

Book design by Christopher Jones

Contents

Preface

I have written this book for the purpose of pointing out
the brutality, futility, and stupidity of war,
especially the tragedy perpetrated by civil war.

The episodes I have described are more poignant messages
against war than the best editorial commentaries
or pacifist preaching.   —*Alexander Riaboff*

*Alexander Riaboff in the cockpit of a Nieuport fighter at the time of his
assignment to the Odessa Advanced Fighter Training School in 1917.*

Editor's Preface

Alexander Riaboff's memoir, *Gatchina Days: Reminiscences of a Russian Pilot,* offers a unique glimpse into the turbulent days of the Russian Revolution and civil war. Written from the perspective of a young military pilot, the memoir blends elements of a personal history and a grim chronicle of a nation caught up in the vortex of war and revolution.

At the core of *Gatchina Days* is Riaboff's personal history of the Russian Revolution. This narrative, written in English, was completed in 1980 after several years of work. His original manuscript has been edited slightly to begin the narrative in 1914, the year in which the nineteen-year-old Riaboff finished secondary school and Russia found itself caught in the Great War. Coverage of Riaboff's early life is incorporated into the introductory essay. The Riaboff narrative covers in chronological order the war years, the revolutionary events of 1917, and the brutal years of the civil war, concluding with Riaboff's escape to China in 1920.

Combined with the memoir proper are supplementary sections: notes to accompany the text, excerpts translated from Riaboff's diary, sidebars dealing with selected themes, an Epilogue, and a brief English-language bibliography. The Introduction is designed to familiarize the reader with the historical context of the Russian Revolution and the less familiar contours of the prerevolutionary tsarist aviation establishment. For clarity, all dates in the memoir follow the Western, or Gregorian, calendar. The old Russian, or Julian calendar, which was thirteen days behind the Western calendar in the twentieth century, was abandoned during the Russian Revolution.

For visual reference, *Gatchina Days* is richly enhanced with a rare collection of photographs, many taken by the author during the period from 1916 to 1923. These rare photographs provide fleeting images of his family, his early training at Gatchina and Odessa, and his military service in Siberia during the Revolution and civil war. Additional photographs have been added to supplement the Riaboff collection.

Several individuals played key roles in preparing *Gatchina Days* for publication. Alice Riaboff, the late Alexander Riaboff's wife, provided valued assistance to her husband in the preparation of the original manuscript. Helen Riaboff Whiteley, the author's daughter, performed a similar role, answering questions about her father's career, supplying relevant portions of his diary for translation, and interviewing surviving family members for additional historical detail. On the latter point, Tatiana Riaboff Hitoon, Alexander Riaboff's younger sister, also responded to many queries about the Riaboff family in prerevolutionary Russia. All of the above were generous with their time and unfailingly responsive to the editor's requests for information.

The staff of the Aeronautics Department of the National Air and Space Museum (NASM) made frequent and substantial contributions to the editorial process. Curators Tom D. Crouch, R.E.G. Davies, and Dominick A. Pisano provided general advice on the organization of the book. Rostyslav Serofimovych, a Smithsonian volunteer working as a translater of Russian language materials, contributed in manifold ways to the project. His expert work as a translator, and his knowledge of early Russian aeronautics, greatly facilitated the preparation of the introductory essay, notes, and diary entries. Other important staff contributions were made by Sybil Descheemaeker, Elaine Fields, and Anita Mason. Dale Hrabak and his staff in the Photographic lab at the National Air and Space Museum supplied professional advice and work in preparing the many illustrations for the book. Efrain Ortiz, the Museum's Office Automation Manager, assisted in many ways in the preparation of the edited manuscript. Frank Pietropaoli and Mary Pavlovich of the NASM Library also provided valuable assistance to the editor in locating varied English- and Russian-language references on the historical period covered in Riaboff's memoir. Behind all these efforts were the constant support and encouragement of Walter J. Boyne, director of the National Air and Space Museum; Donald S. Lopez, deputy director; and E.T. Wooldridge, chairman of the Aeronautics Department.

—Von Hardesty
Washington, D.C.
September, 1985

Introduction
by Von Hardesty

When Alexander Riaboff graduated from the Gatchina Military Flying School in January 1917, Nicholas II still occupied his shaky throne and Russia was at war with Germany and the other Central Powers. Riaboff and the graduating cadets at Gatchina, newly commissioned as ensigns, entered the Imperial Russian Air Force (IRAF) as military pilots. For Nicholas II, the signing of the formal papers approving the military commissions at Gatchina was one of his last official acts as Emperor of Russia. Within weeks Russia found itself caught up in an epic revolution, a cataclysm that toppled the Romanov dynasty after three hundred years of rule and propelled Russia down an uncertain path toward revolution and civil war.

The revolutionary events of 1917—mutinies at the front, rural uprisings, and workers' strikes in the major cities—quickly shattered the Old Regime's fragile hold on authority. Nicholas II finally abdicated in March of that year. A Provisional Government inherited formal power and vainly sought to keep Russia in World War I and to forge a new constitutional order. Rival foci of power, the soviets, or councils, composed of radical elements drawn from the armed forces and the urban industrial districts, simultaneously challenged the Provisional Government for leadership. By November 1917, the Bolshevik party under Lenin had seized control and moved aggressively to consolidate its monopoly of power. Various political factions across the vast domain of the former Russian Empire in turn arose to oppose the Bolsheviks. This brutal struggle, a civil war that endured until 1921, brought triumph for Lenin and his Bolsheviks (now called Communists), but only after great devastation and loss of human life.

Alexander Riaboff in his memoir, *Gatchina Days*, provides a unique historical account of these years of war and revolution, as seen through the eyes of a young Russian pilot. Riaboff's personal saga as a military pilot included military service under several banners: a brief tour of duty

Ye. B. Rudnev, pioneer Russian aviator, flies over St. Isaac's Cathedral, St. Petersburg (now Leningrad), in September 1910, at the time of the first All-Russian Festival of Aeronautics.

Alexander Riaboff as an engineering student at the Institute of Communications, Moscow 1914.

in the Imperial Russian Air Force, an interlude of flying under the formal authority of the Provisional Government, an unpleasant and dangerous stint with Lenin's new Red Air Fleet, and, finally, defection to the Whites (as the anti-Bolshevik elements became known) for two years of military flying in the doomed effort to topple the Bolshevik regime. Riaboff found himself buffeted by the swift and ever-shifting forces of revolution and counterrevolution during these turbulent years. Despite the hardships, the constant movement, and the austere conditions imposed on the small air units that operated across European Russia and Siberia, Riaboff was able to fly and to fly often.

Through Riaboff's memoir we see the demise of the old tsarist military aviation establishment; glimpses of the embryonic Red Air Fleet, the forerunner of the modern Soviet Air Force; and images of Riaboff's flying career with the ragtag units of the Whites fighting the Bolsheviks across the vast expanse of Siberia. Riaboff illustrates in *Gatchina Days* the nearly forgotten fact that aviation played a sustained, if limited, role during the Russian civil war. On a more personal level, Riaboff's life, as recounted in *Gatchina Days*, conveys to the modern reader a compelling image of human decency, personal sacrifice, and heroism in a time of great social upheaval.

The triumph of the Bolsheviks in the Russian civil war forced Riaboff into exile in 1920. Four long decades passed before he again, as an

American citizen, visited his former homeland. On this visit to the Soviet Union in 1960, Riaboff was reunited with surviving members of his family. Moreover, he discovered that they had preserved many of the photographs he had taken as a cadet at Gatchina Military Flying School and later as a military pilot during the years of the Revolution.

This rare collection of photographs, showing nearly forgotten scenes of aviation life in revolutionary Russia, prompted a determination in Riaboff to write his memoirs, a task he only completed in 1980. He was motivated by his desire to make a statement about the tragic consequences of war, especially civil war, and about the darker side of a revolution ostensibly dedicated to liberation. He hoped that his experiences would lead to an understanding of a larger truth about the destructiveness of war and revolution.

At the same time, this rich legacy of written and photographic materials allows us to see, if through a glass darkly, the lost world of early Russian aviation. Riaboff captures for us his involvement in Russian military aviation from Gatchina to Odessa, and then, following an eastward trek, to Moscow, to Kazan, and, finally, to the improvised airstrips across Siberia in the brutal civil war.

One of the most remarkable aspects of the Riaboff memoir is the absence of any narrow political partisanship that so often is characteristic of Russian émigré literature. He did not use his personal recollections as an occasion merely to attack the Bolsheviks. Riaboff brought to his writing an attitude of reflection and objectivity, rarely bitterness or recrimination. He saw his own survival as remarkable and viewed the events following the Russian Revolution as a modern time of troubles for Russia. The tragic aspects of this revolutionary period never escaped his consciousness. In retrospect, he saw his own personal ordeal as representative; countless and nameless fellow countrymen endured the same hardships.

As a youth, Riaboff's political outlook was devoid of political extremism. In fact, he displayed a general indifference to political issues. In an age when students were often highly politicized, Riaboff displayed a narrow and highly focused interest in his studies. While in the Russian Army in 1917, he expressed few regrets for the passing of the Romanov dynasty, and in the violent period that followed showed little sympathy for the monarchist cause. Toward the Bolsheviks, however, Riaboff gradually became an ardent foe. Bolshevik rule sharpened his political consciousness, even as it threatened him and his family. He had observed Bolshevik rule firsthand and criticized it for its gross inhumanity and ruthlessness. As time passed, Riaboff measured the Bolshevik cause on how it actually functioned, not on its self-serving party slogans. For these reasons, he felt the Whites were infinitely better as a political alternative than Lenin's disciplined revolutionaries. Despite Riaboff's frequent hardships, dangers, and personal disillusionment with many of his commanders, he remained active in the White struggle against the Bolsheviks until the very end.

Riaboff, a self-described "liberal," no doubt would have preferred some form of democratic rule for post revolutionary Russia, although his own political ideas, as he admits in his memoir, developed slowly during the period of the civil war. Earlier, one of Riaboff's secondary school teachers, a man named Il'ya Malakhov, had made a deep impression. Malakhov had argued fervently that governmental structures and political parties were less important than the personal character of those in authority, the notion that the sources of poor government rested with human nature not the political structure. Riaboff carried this nonideological approach with him throughout the years of the Revolution. His commitment to the Whites was enduring, but his allegiance was always tempered by an abhorrence of the occasional brutality he observed among White army units and his criticism of the Whites for their failure to build a viable political program. Riaboff saw in these two weaknesses the seedbed of the ultimate Bolshevik triumph.

Born on April 23, 1895, Alexander Riaboff came from what one might describe as a "middle class" family in the context of the late tsarist period. His father, Vassiliy Riaboff, was a manager of a woolen manufacturing firm in Moscow. His strong-willed mother, Eudoxia, made a powerful impression on the Riaboff family. She organized the family finances to build a dacha, or country home, at Chulkovo, outside Moscow. For Alexander and his three younger sisters, Anna, Panya, and Tatiana, their visits to this family dacha became a pleasant and enduring memory of childhood.

The ownership of a dacha, of course, suggested that the Riaboffs, if not wealthy, possessed considerable financial means as members of Moscow's growing industrial class in prerevolutionary Russia. The family was hardworking and conventional in all respects, except for one uncle, Stepan Riaboff, who in the years before 1917 was an avowed revolutionary.

As a successful plant manager, Vassiliy Riaboff took steps to assure that his bright and highly motivated son obtained a proper education. To this end, the young Riaboff was enrolled in the Alexander III Commercial School, one of Moscow's excellent preparatory schools offering a highly structured eight-year course of study. Riaboff responded positively to the Alexander School, which trained the children of Moscow's commercial class and was known for its strict discipline and academic standards. When he graduated in 1914, Riaboff received a gold medal for scholarship. His academic record showed an aptitude for mathematics, science, and foreign languages. Riaboff's teachers encouraged him to enter an engineering program at the Imperial Naval Academy in St. Petersburg. He rejected this advice, choosing instead to enter the Institute of Communications, which offered a program of engineering closer to home.

The Riaboff family in 1897. Vassiliy Ivanovich and Eudoxia Vasiliyevna Riaboff pose with their two children, Alexander, aged two, and Anna, aged six months.

The advent of the Great War in 1914 did not at first interrupt his education plans, but as time passed Riaboff found it more difficult to stay at the institute. By 1916 the war crisis had deepened, and, as a consequence, the tsarist government ordered a military draft that soon reached into the schools and universities. Riaboff had been patriotic during these first two years of the war, and increasingly he had viewed his civilian status as a source of embarrassment. Faced with the external threat of the draft and the internal urgings of patriotism, Riaboff decided to act on his own, to shape his own destiny. His decision, after a careful assessment of the options, was to volunteer for the Imperial Russian Air Force (IRAF).

Riaboff came to aviation belatedly, almost by accident. His initial interest in flying was war-induced; he saw in the IRAF a romantic alternative to the drudgery of the infantry or the boredom so evident to him in the other branches of the armed forces. When recounting his happy days before 1914, Riaboff makes no mention of any fascination with the airplane or memories of early Russian aeronautical life.

This absence of comment on aviation is interesting because Riaboff's native Moscow in the prewar years was a major center for aeronautical activity. The Moscow Aeronautical Society, affiliated with the Imperial All-Russian Aero Club, sponsored many air meets and aircraft exhibitions during Riaboff's youth.

Moscow also took pride in having the Duks aircraft factory, the largest in Russia. In 1911, for example, Russia followed intently the dramatic St. Petersburg-to-Moscow air race, which covered a course of 400 miles. When A. A. Vasil'yev landed at Khodinskoye field in his Bleriot, the only competitor to make the arduous flight over the forests of northern Russia, all of Moscow, it seemed, rushed to greet him. Curiously, Riaboff makes no mention of this event or of any youthful interest in aviation.

Flight training for Riaboff began in Moscow in 1916, at the School of Theoretical Aviation, where Russia's famed aerodynamist Nicholas Ye. Zhukovskiy was one of his instructors. A talented engineering student, Riaboff quickly mastered the theoretical side of his training. His encounter with flight training at this stage appears almost casual and bemused. After four months of studying flying in the abstract, Riaboff was sent to Russia's largest and best-equipped training facility, the Gatchina Military Flying School located twenty-five miles outside Petrograd (formerly St. Petersburg). Throughout the practical phase of flight training, the young cadet displayed an aptitude and a growing enthusiasm for flying, always confident of his own abilities and critical of those who approached flying with fear or recklessness. Riaboff developed a keen interest in aviation during this period, but at no time did flying become a pursuit that dominated his life.

The IRAF that Riaboff entered in 1916 was small by comparison with

Igor Sikorsky's famed four-engine "Grand," which first flew in 1913. This remarkable aircraft reflected Russia's pioneering skill in building large multi-engined flying machines.

Russia held its first International Exhibition of Aeronautics in St. Petersburg in 1911.

The Grand Duke Alexander with his three sons at the All-Russian Festival of Aeronautics in 1910. As a member of the Romanov dynasty and a high-ranking naval officer, the grand duke did much to promote aviation in prerevolutionary Russia.

the air forces operating on the western front, and it faced the enormous challenge of maintaining itself in a context of military defeat. Russia's industrial base for aviation was never large and the acute shortage of aero engines undermined the nation's war effort in the air. Among the armed forces of the Russian Empire, perhaps the IRAF possessed the highest morale. It was an elite service, noted for its élan and bravery. One of the popular heroes of the war was Peter Nesterov, who downed an Austrian airplane in August 1914 by the deliberate use of the tactic of ramming. There were other effective fighter pilots in the IRAF, for example, Ivan Kruten and A. A. Kazakov, who drew considerable public attention.

When Riaboff arrived at Gatchina in 1916, the Russian War Ministry had just instituted a new flight-training curriculum that aimed at providing practical instruction for Russia's growing pool of military pilots. The training program was well conceived and provided thorough work in both theoretical and practical phases. The war had brought considerable attrition to Russia's small cadre of military pilots, and the wartime uses of aviation had exposed the need for a more intense program of training by cadets at Gatchina. Riaboff fell under this reformed program of instruction and his memoir suggests that the Gatchina school was a demanding place, requiring the cadets to pass through a gauntlet of tests. There were numerous washouts as the instructors identified the most proficient cadets for acceptance into the air force.

The IRAF operated twelve flight-training schools by 1916, with Gatchina and the large Sevastopol Flying School in the Crimea as the two major facilities. Training was offered for pilots, pilot-observers, and engine mechanics. Among the three, the pilot-observers began to assume greater importance as Russian military aviation, in the absence of much air-to-air combat, was viewed as an effective tool to assist artillery units and to provide systematic aerial reconnaissance. Riaboff was trained as a military pilot, and after his stint at Gatchina, was posted to Odessa where the IRAF operated its advanced fighter-training school. All fighter pilots, by virtue of the 1916 training program, were required to learn the techniques of aerial reconnaissance. Competent engine mechanics were difficult to mobilize and to train in large numbers. For this reason a substantial percentage of IRAF engine mechanics were POWs, offered service in Russian air units as a way to avoid the rigors of the prisoner-of-war camps.

Alexander Riaboff gives one fleeting glimpse of the IRAF's remarkable Il'ya Muromets, the world's first four-engine bomber/reconnaissance aircraft. He took one photograph of an Il'ya Muromets at Moscow in 1917. The Murometsy, as the multi-engined aircraft were called, provided the IRAF with unique long-range flying fortresses that performed myriad bombing and reconnaissance missions in the war. Igor Sikorsky had designed the enormous aircraft in the prewar years and had made a number of dramatic flights, including one epic round trip flight from St. Petersburg to Kiev (1,800 miles) in 1914.

While Riaboff did not comment in his memoir on the Il'ya Muromets, it is important to note that his photograph provides further evidence that the Sikorsky's Murometsy, organized during the war into a special flying corps, maintained an active role throughout the period Russia was in World War I. Later, the Bolsheviks dragooned the Il'ya Muromets into service as part of the Red Air Fleet. From Riaboff's coverage of the civil war period, it is apparent that both the Bolsheviks and their opponents, the Whites, made extensive use of tsarist aircraft and equipment.

The bulk of the Riaboff memoir covers the Russian civil war that raged from 1918 to 1921. His coverage of these turbulent years is done in broad strokes, often leaving critical events and personalities mentioned but not always discussed in detail. For this reason, his diary has been translated and excerpts have been incorporated into *Gatchina Days* to provide additional information. One is impressed with the image Riaboff leaves us of aviation in Siberia during the civil war— the austere conditions and equipment shortages, the relative freedom of military pilots to move from one squadron to the next, the limited numbers of aircraft operating over vast distances, the frequent movement of the participants, and the indescribable brutality of the conflict.

Peter Nesterov stands next to a French-made Morane monoplane. Nesterov performed a loop in 1913, the first executed successfully. He died in aerial combat while ramming an Austrian airplane over Galicia in August 1914.

Alexander Riaboff's memoir reflects in its historical detail the intensity of the Russian Revolution. It is filled with perpetual upheaval and movement, spirals of violence, and exacting retribution. Politics for Russia in this period, as George Kennan has described it, was a "mortal exercise." Mutual tolerance or trust in the motives of one's political opponents was nonexistent. Most political groups shunned moderation and, if they shared any trait in common, it was the disdain for any form of political accommodation. A fierce embrace of ideology or the defense of vested interests precluded compromise. All political opponents were viewed as misguided, as enemies of truth; if allowed to obtain power, they would bring ruin to society. Each side approached the civil war in absolute terms, seeing in the conflict a life and death struggle in which the primitive application of force was the sole means of survival. Against this backdrop, Riaboff is a rare figure who followed his own sense of duty outside the rigidities of an ideology and with a remarkable sense of toleration.

The locale for Riaboff's experiences is the vast expanse of Russia and Siberia. Between 1918 and 1920 he moved eastward across the forests and steppes of Siberia with the long and winding Trans-Siberian Railroad as the sole means of defining position and direction. During these years the rugged Siberian frontier became the arena for Reds and Whites to engage in a wild and confusing battle for survival.

Along the narrow zone of the railway and the towns that punctuated the endless Siberian landscape, the armies and partisan bands engaged in a barbaric struggle. There were alternating Red and White terrors, random killings, mutilations, and pillaging. Partisan bands of various

Igor Sikorsky built the Il'ya Muromets as a successor to the Grand. Here a military version is parked at a frontline airfield during World War I.

political hues, bandits, and assorted armed groups swept back and forth across the chaotic passageway of the railway. Mixed into this cauldron of violence were the Czech Legionaires, former prisoners of war trying to escape to the West, and the Allied interventionists (British, French, American, and Japanese contingents and agents), giving general support to the anti-Bolshevik cause and, in the case of the Japanese, actively bankrolling bandit groups thinly disguised as armies.

In the cities there were food shortages, political intrigue, and frequent uprisings. As the Red Army pushed aggressively eastward, countless refugees staggered ahead of them, facing starvation, the extremes of the Siberian climate, and the threat of pandemic diseases such as typhus. Siberia found itself in 1919 and 1920 gripped by near anarchy as various groups jockeyed for position and the disillusioned population prepared for the inevitable triumph of the Bolsheviks.

Riaboff represents that element of the White armies characterized by integrity and honor. Too often these officers were compelled to serve under commanders unworthy of their trust. These rapacious and corrupt White generals frequently and, in the last analysis, fatally, compromised the anti-Bolshevik cause. Riaboff's memoir and diary mirror this reality in vivid and melancholy detail. The Bolsheviks won the prolonged struggle for Siberia and ultimately for Russia itself because they displayed organization, discipline, and unrivaled skills as propagandists, the very traits the Whites could never muster in any sustained fashion.

Alexander Riaboff in his memoir *Gatchina Days* recreates for us this drama of war and revolution, as he experienced it during his youth as a Russian military pilot.

1. My Decision to Become a Pilot

Alexander Riaboff begins his memoir in the fateful summer of 1914, at the time Russia entered World War I. The beginning of the war followed a few weeks after his graduation from the Alexander III Commercial School in Moscow. At the time, he was preparing to pursue a career as an engineer.

The gallant efforts of the Russian Army to achieve a decisive victory over Germany in 1914 had proven ineffectual. At the Battle of Tannenberg, the German Army scored a major victory over the Russians, forcing Nicholas II's army into retreat. The year 1915 brought further defeats for Russia at the hands of the Central Powers, despite the fact that Nicholas II had assumed personal command of his troops. By 1916, Russia faced a severe political crisis brought on by military defeat and a deteriorating economic situation. At the top was the emperor, discredited and threatened by an elemental revolution from below.

During the first two years of the war, Riaboff had remained in engineering school, but he found himself increasingly restive in civilian life. Finally, he joined the Imperial Russian Air Force in 1916. His enlistment led to flight training at the Gatchina Military Flying School. Even as Riaboff arrived at Gatchina, the first rumblings of revolution were apparent in Petrograd, a mere twenty-five miles away.

■   ■   ■   ■

WHEN WORLD WAR I BROKE OUT, I was a nineteen-year-old youth living with my family in Moscow and preparing to take the entrance exams at the Institute of Communications. The morning of August 1, 1914, remains vivid in my mind. I had gone out early to buy groceries for breakfast. My three sisters and mother were vacationing in Chulkovo at our dacha twenty-five miles outside the city. My father and I had remained in Moscow during these sweltering summer days.

As I strolled along, my mind cluttered with examination minutiae, my eyes suddenly fixed on a newspaper headline: GERMANY DECLARES

Alexander Riaboff (center) at the time of his application to the Institute of Communications in the summer of 1914.

WAR ON RUSSIA! Seconds passed as I stood there staring at these ominous words, a pandemonium of random thoughts racing through my mind. Finally, the implications of these headlines struck me: "What am I going to do? How about my exams? Will they be cancelled? Will my uncles be drafted? Will *I* be drafted?"

Quickly my thoughts focused on my personal lot and the implications of the war on my future life. How would the war disrupt my plans to become an engineer? Here I was, ready to embark upon a new stage in my life, and now events quite out of my control threatened to disrupt my plans entirely. I bought a newspaper and ran home to my father, Vassiliy Riaboff, who was a manager with a woolen manufacturing firm in Moscow. We had discussed the possibility of war with Germany. In fact, this question had been discussed frequently in our home, but the actuality of war came as a genuine shock to him.

My mood was one of utter despair. At this stage my reaction was personal, not patriotic in the conventional sense. I turned to my father and said: "Isn't it ironic that we, the young and the most fit, are required to fight. Why must we suffer the agonies of war while the instigators only play a game of maneuvers?" To comfort me, he said that the war might be short-lived—"all we can do is wait and hope."

My father's attitude was typical. Few Russians in the summer of 1914 expected a four-year war of attrition and military defeat, let alone a massive social and political revolution. We failed to anticipate the course of history for the next few years. We did not realize that the war, as a sledge hammer, would shatter our lives in a profound and irreversible fashion.

Just two months had passed since I graduated at the top of my class from the Alexander Commercial School in Moscow, earning the gold medal for scholarship. At the urging of my teachers and family, I had dedicated myself during the summer of 1914 to a rigorous course of study to prepare for the entrance examination at the Institute of Communications. My father was determined that I should continue with these preparations to enter the institute. He urged me to take the exams and proceed with my studies.

My friend Sergei Popov had joined me in this arduous quest. We both had visions of becoming engineers. The preparatory work consumed much of our time. We memorized hundreds of theorems with proofs and applications. My personal involvement in the study of mathematics became nearly complete, making me almost oblivious to the outside world and the dramatic events unfolding around me.

On the day of the examination I joined 550 applicants drawn from different regions of the Russian Empire. Along with Sergei Popov, there were several friends from Moscow who decided to take the formidable exams. All applicants came well prepared; most had received high marks in mathematics and physics in their respective schools. We were divided into groups of twelve and required to follow a precise schedule of testing. The examination process was grueling and offered little solace to those unprepared. Every applicant had an examination book, which contained his identification photograph, his signature, and his schedule. At each stage of the testing process the student presented his book to the examining professor, who first carefully verified his identification. The professor then gave the student a ticket or a folded piece of paper that indicated the theorem to prove and/or the mathematical problem to solve.

At a blackboard in full view of the examining professors and the other applicants awaiting their turns, each aspiring student faced his moment of truth. If an applicant passed his examination, the professor entered his grade and returned the book to him. If he failed, the examining professor retained the book, which signalled in an abrupt fashion the end of the exams and one's hopes of entering the Institute of Communications. Trigonometry, as it turned out, proved to be the most devastating part of the test, shattering many dreams on that day.

Sergei and I passed the exams. Once the results were disclosed, it was discovered that many applicants had failed; only 118 had passed. This still created a problem because the institute was permitted to accept only 100 students. How was this problem to be solved? After some deliberation, it was decided to appeal to the Emperor, Nicholas II, to seek his permission for the acceptance of all the applicants who had passed the exams. A telegram was duly sent to St. Petersburg. Permission was granted.

The war, however, quickly altered our lives. For a while, there was a period of normalcy. I proceeded to study diligently at the Institute of Communication, hoping to complete my course of studies as soon as possible. In addition, I worked as a tutor at the institute. This intensive schedule kept me fully occupied, with little time for relaxation or entertainment. After all, my country was at war! I felt I should be as diligent as possible. Events soon caught up with us. Members of my family found themselves drawn into the conflict. My uncles, Ivan and Pavel, were called up from the reserves and were sent to the front. Some of my friends were drafted and others volunteered.

I felt uncomfortable. My studies were important and all-consuming, but there was a feeling of uneasiness with my student status. I felt it my duty to help in my country's difficult struggle in some way. As the months passed, volunteers were needed to unload and reload the returning wounded into buses and trucks from the troop trains that reached Moscow. Many of the war's casualties were brought to the city every night by rail, either in transit to other cities or to be sent to one of Moscow's hospitals. I offered my services, and this experience became my first occasion to view up close the horrors of war.

Thousands of wounded were delivered to us night after night—their bandages bloodied, uniforms dirty and torn, bodies pitifully mutilated—and most of them suffered silently. There were the hopeless cases, the dying and the starving, and always the grim and sightless eyes of the dead. Others were barely ambulatory, limping painfully or trying to move about on makeshift crutches; some supported badly wounded arms and hands while they waited patiently for medical help; many were transported on stretchers.

I remember the outburst of one wounded soldier: "It's tough, bad for us, but it's just as bad and tough for the Germans and Austrians: We needed guns, rifles, and ammunition. . . . you can't fight a war without 'em . . . God help us!"

The war was fought on relentlessly, and without a quick victory, pulling more and more Russian youth into the bitter conflict. Between 1914 and 1916, the number of students at the institute diminished rapidly: many had enlisted and a few had gone to work in the war industries. One day I saw an advertisement posted at the institute calling for students to work as ammunition inspectors in one of Moscow's munitions factories. I decided that this kind of work would be an excellent way to make a solid contribution to the war effort. During the winter of 1915, I became a hand-grenade inspector along with three other chaps, one of whom was a law student at Moscow University.

One day this law student came to work late in the morning, looking somewhat worried. Some time passed, and then he called me aside. "Say Riaboff," he whispered to me, "would you mind stepping outside the shop to see if you can spot a short fellow in civilian clothes?" I obliged him, took a look in the specified direction for the suspicious person, and reported: "Yes, I saw your man and another fellow with him standing at the corner of the block."

Without hesitating, the student hurriedly put on his coat and hat and carefully slipped away through the back door of the factory. He obviously was greatly perturbed. Shortly thereafter, the two men, accompanied by the police, entered the shop in search of the student. After a careful inspection of the shop's occupants, they left without their victim. It was then apparent to me that my fellow worker was a revolutionary and the tsar's secret police, the Okhrana, had just missed their opportunity to apprehend him. This incident brought home to me the specter of revolution that haunted Russia during those days.

Two years passed. Each day brought me closer and closer to the realization that I would be drawn directly into the war, the emerging

When World War I began, the tsarist government quickly mobilized its small aviation industry to build military aircraft. The Duks plant in Moscow, Russia's largest aircraft factory at the time, produced a large number of aircraft for the war effort.

revolution, or both. At the beginning of 1916, there was no doubt in my mind that the war would continue for a long time. While my good academic record made me temporarily immune to the draft, there was still the possibility of induction if the war emergency worsened. At the same time, the wartime propaganda with its patriotic appeals was taking its toll on me. With each passing month, I became more uncomfortable with my civilian status. I feared being perceived as a slacker. When the pressure to join became too severe for me to withstand, I decided to volunteer while I still had the opportunity to choose the branch of service I wanted to join. I realized I had to make a very important decision in selecting a particular branch most suited to me.

I was well aware of the average life expectancy of army officers at the front in 1916. Attrition was high and Russia was losing the war. In the infantry I would probably remain alive for three months, and in the artillery, perhaps for six months. I was not interested in joining the cavalry or the engineering corps. So I plotted another course of action. Being an individualist and preferring to be killed outright rather than be crippled

The fuselage-testing lab at the Moscow School of Theoretical Aviation. Alexander Riaboff (seated, left) began his flight training at this school in 1916.

physically or mentally, I decided to become an air force pilot. My recent experiences in transferring the maimed, wounded, and dead troops had remained vivid in my memory. I preferred the lot of military pilots, even if the risks and dangers were high.

I confided my decision to my friend Sergei and asked him what I should do. "You are wrong, absolutely wrong!" he said vehemently. "Don't you remember how difficult it was to get into engineering school? How hard we worked to pass the examinations? Now we have been studying a year and a half and we are both doing fine. We have passed the necessary examinations and they cannot conscript us. You want to go to war to save Russia! War, as you must know, is sheer insanity. It means murder, destruction, and endless suffering. Some day the war will end and life will return to normal, and that will be the time Russia will need us—engineers to rebuild her."

He paused a moment, then continued pleadingly, "Stay with me and finish your studies. Stay out of this insanity. You may be killed—

needlessly, stupidly, and without having done anyone any good. And get this straight—once you quit your studies you will have one hell of a time starting all over again! Two or three years away from school and you will have forgotten all you have studied and then you'll have to start from the beginning. I'll bet if you go to war you'll never return and all the work, studying, and knowledge gained will have been for nothing." He turned and looked me straight in the eye: "As far as I am concerned, I am going to do everything in my power to stay away from war and finish school at any price!"

Despite his pleas, I held firm to my decision to leave school. Sergei's aim to remain a civilian turned out to be difficult. He eventually was called to military service, but he resorted to every kind of trick to avoid the draft. Finally, he was able to graduate from the institute. He subsequently became an engineer and worked in that capacity for many years. My family kept in touch with him, as he visited them regularly, but then he stopped coming and they lost track of him completely. One of the great losses of these difficult years was my contact with old friends who were killed or separated from me.

Before I proceeded with my plan to join the military and contribute to the war effort as a pilot, I faced the task of telling my parents of my plans. One evening, early in 1916, during a family dinner, I said to my father, mother, and sisters that the war would continue for a long time and most assuredly I would be called to serve. "If I volunteer," I told them, "I'll be able to choose the service I prefer."

Dead silence followed. Father then asked: "What do you have in mind? Where would you like to serve?" "I want to be a flier," I declared. My announcement was a bombshell to my parents. "Why, son, do you want to join the air force?" my father implored. He knew the dangers of flying, especially in the army. "The truth is," I answered, "I don't want to be crippled for life. I prefer to be killed outright. Relatively few fliers are maimed in crashes—most of them are killed—instantly."

This graphic statement spoke a harsh truth, but did little to reassure my parents. Mother's eyes filled with tears, my sisters' faces expressed horror and disbelief, but Father summed up the conversation by stating as calmly as his emotions permitted by stating: "This is your life and your decision to make. Do whatever you think is best for you. God help you, my son!"

Now I faced the challenge of making my goal a reality. Shortly after my conversation with my parents, I saw a notice posted at the Institute of Communications inviting students to enter the aviation school in Moscow. Volunteers would have to take a three months' course on the theory of aviation and then they would be sent to the Gatchina Military Flying

School near Petrograd for another three months of flight instructions. Graduates would then be sent to the front.

Without further delay, I submitted my application, passed the required medical examination, and within a short time reported to the school, which was located close to our home. My class consisted of a group of sixty-three young men, most of them about my age and from engineering schools, although some came from liberal arts schools. One cadet was an attorney and two others had university degrees. We were housed in a large building, a former private residence, surrounded by a big yard and garden. The top floor was occupied by dormitories and the lower floor had been made over into a complex of lecture halls, a dining room, and a library.

A war veteran, Captain Turchinsky, was in charge of the school. A Sergeant Subbotin assisted him. The captain, a regular army officer, had been badly wounded, resulting in his having lost the use of his right arm and leaving him in generally poor health. He understood that we were not an ordinary group of village boys and treated us accordingly with politeness and understanding, although insisting upon some degree of discipline.

The sergeant, a long-time army man, also understood that we were of the "intelligentsia," as he referred to us, but from force of habit he required a much stricter enforcement of discipline. We adjusted to our new way of life quickly without incident. After eating a good breakfast each day, we went through a course of regular army drills. Soon we were able to march, turn, and salute easily and smartly. Some of the fellows were obviously uncoordinated in their movements and later proved to be unsuited for flying.

After our drill exercises, we marched to the Moscow Technological Institute, located a short distance from our quarters. This school had an aviation department headed by Professor Nicholas Zhukovskiy, then Russia's foremost aerodynamicist. The school was equipped with a very large wind tunnel. We worked with the wind tunnel and attended lectures by Professor Zhukovskiy and his staff. Although he was elderly, over seventy at the time, he impressed us greatly with his complete dedication to his task.

Typically, as Professor Zhukovskiy lectured, he became so engrossed in his subject that he appeared to be in another world. He was also well known for his absent-mindedness, and there were stories pertaining to his eccentricity. It seems that once Zhukovskiy went to his own home, pressed the door bell, and waited. His housemaid opened the door and, seeing her master standing there with a far-away look, said, "What do you want sir?" The professor inquired: "Is the master of the house in? I would like to see him."

Nicholas Zhukovskiy in his Moscow study. The famed Russian aerodynamicist gave lectures at the Moscow School of Theoretical Aviation when Riaboff attended the primary flight school in 1916.

"No sir, he is not in," responded his faithful servant.

"Oh well, that's all right. I'll return later," and he ambled down the street unaware of his disorientation.

It was one of the customs at the school for the cadets to rise when any professor entered the classroom and to remain standing until we were told to be seated, usually when the professor reached the rostrum. Zhukovskiy, however, sometimes went to the rostrum and proceeded to lecture without noticing that we were standing, motionless and unable to take notes. In due time we became aware of his propensity for looking at us with unseeing eyes and, after waiting several minutes, we simply seated ourselves without waiting for his command. Many times I wondered if he ever noticed our being there at all.

Sixty-three of us graduated from the Moscow School of Theoretical Aviation. Upon graduation, it was exciting to hear that we had been assigned to the Gatchina Military Flying School. We took the long train ride from Moscow to Petrograd, a 400-mile journey across northern Russia. Gatchina had been a major air park before the war, and now it housed a large training school for military pilots. The flight school was adjacent to the Gatchina Palace and the town of Gatchina, which was located twenty-five miles outside Petrograd, on an important railroad line connecting the capital with the western provinces of the Russian Empire.

2. Gatchina Military Flying School

The Gatchina Military Flying School in 1916 was the chief training center for pilots in the Imperial Russian Air Force. Located on a vast air park near the Gatchina Palace, once the home of Romanov emperors, the school owed its inception to the Grand Duke Alexander, cousin of Nicholas II, who in the prewar years had been an ardent advocate of air power through his "Committee for the Strengthening of the Air Fleet." The grand duke's efforts had led as well to the founding of a year-round flight training facility at Sevastopol on the Black Sea.

Gatchina's large grass fields allowed cadets to practice takeoffs and landings simultanously on two separate, but connected, airstrips. There were a number of permanent hangars, converted barns used for repair shops and storage, as well as tent hangars. The flight-training program at Gatchina was divided into primary and advanced phases. Cadets did not train as classes, but were admitted to the program of flight training as vacancies occurred. There were approximately 200 aircraft trainers at Gatchina, mostly French-designed Farman 4s and Farman 16s built under license by the Russians. Many of the mechanics and technicians at Gatchina were Austrian prisoners of war.

The Farman 4 was the primary trainer. This aircraft was simple in design, rugged, and easy to maintain, although in 1916 a rather obsolete type. The Farman 4 was powered by a seven-cylinder Gnome rotary engine, which threw ample doses of castor oil backward over the fabric of the aircraft and was subject to frequent breakdowns. The worst feature of the Farman 4 was its tendency to sideslip in turns, even shallow turns, so there were frequent accidents—as Riaboff soon discovered.

■   ■   ■   ■

Alexander Riaboff, fourth from left on the ground, is pictured with cadets and instructors at Gatchina in 1916.

Riaboff took this aerial shot of the Gatchina Palace and air park while following the railroad tracks, heading northward. The practice field for the flying school is clearly visible in the center, just below the palace grounds.

A close-up shot of the Gatchina Palace and lake. The palace was used in earlier times by the Romanov dynasty.

The Gatchina Military Flying School complex from the air, looking to the northwest.

View of the airfield from the Gatchina side. Rows of soldiers on the drill field are at the right. The tents at lower left, next to the railroad tracks, belong to a battalion of troops assigned to the railway.

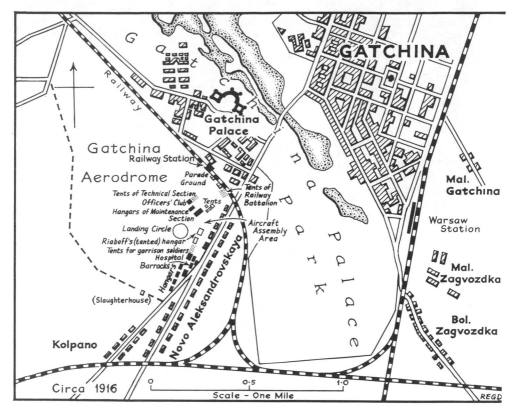

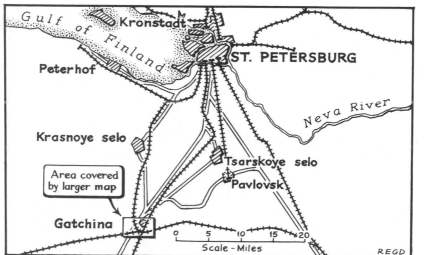

Gatchina

Alexander Riaboff began his flight instruction at the Gatchina Military Flying School in 1916. This plan shows the large air park, located adjacent to the town and the historic Gatchina Palace, and the accompanying map shows its location.

W E MADE THE LONG TRAIN RIDE TO GATCHINA. Upon arriving, we were lined up in front of the school's administration building and were formally greeted by the commander, Colonel Gorshkov, and his staff. Our squadron leader officially reported to the colonel that his group of sixty-three volunteers had arrived from Moscow for flying instruction.

Colonel Gorshkov smiled pleasantly and said, "I don't know what to do with you. We have neither airplanes nor instructors to teach you. As a matter of fact, we don't even have a place for you to sleep! I reported this situation to headquarters and they sent you in spite of my warning. You are the second group to arrive and I have no idea how long it will be before you will be assigned to flying squadrons!" After that disappointing bit of news, he abruptly dismissed us.

That same day, and without any explanation, we were pleasantly surprised that this same commander assigned us to spacious barracks provided with beds and clean linen. Apparently the former occupants of the building had been evicted in order to make room for us. We were served simple but tasty food, and after some interval were assigned to various flying groups—not for the purpose of learning to fly but to learn how to take care of the airplanes and to get acquainted with the school's routine and operations. This experience proved to be useful, as we gradually

When Riaboff first arrived at Gatchina, he was quartered in the summer home of Count Benkendorf, which had been appropriated for military use.

learned about the school, the aircraft, and the technical language of military flying.

At Gatchina the Imperial Russian Air Force used two types of trainers for flight instruction. The Farman 4, an obsolete prewar biplane, was used for the primary phases of training. It had an open seat for the pilot on the leading edge of the lower wing, which permitted him to see the ground directly below. It had a light wooden frame extending forward from the edge of the lower wing carrying a pivoted crossbar, which operated the rudder at the rear end of the aircraft. The crossbar was operated by the pilot's feet: you pushed with the left foot to turn the machine to the left and pushed with the right foot to turn it to the right. A stick, connected to the ailerons, was used to tilt the trainer by pushing it respectively to the left or to the right. The stick was also connected to the elevators, so that by pushing it forward or backward you could move the nose of the aircraft down or up. The durable Farman 4 had a fifty-horsepower rotary engine and could reach a top speed of about thirty-five miles per hour.

In time, the school obtained the Farman 16, which was a great improvement over the Farman 4. The Farman 16 had a closed fuselage with an open cockpit for the pilot, and behind him there was a passenger compartment. With a rotary, eighty-horsepower engine, the aircraft was more reliable and easier to handle. Its maximum speed was about sixty miles per hour. We also flew Farman 21s at Gatchina.

While in training, cadets were gradually assigned to groups using one or the other of these trainers. I began my lessons in a Farman 4. My first lessons were confined to steering the trainer on the ground along a lengthy strip separated from the airfield by a narrow grove of tall trees and bordered by a railroad on the other side. To keep this clumsy old airplane on a straight course proved to be difficult, especially in a cross wind. Each of us took some time to become proficient at this maneuver.

One day during my training, before I had been permitted to fly, an unusual incident occurred. I had taxied the Farman 4 to the end of the strip, turned it around, and was returning to the starting point when I suddenly spotted another Farman 4 emerging through the passage from the field to the strip and directly in my path. The speed of my Farman and the distance between the two trainers were such that a collision seemed unavoidable. Without hesitation I opened the engine throttle fully, pulled the stick back, and suddenly found myself airborne. My original idea was to leap over the other airplane and land as fast as possible. However, when I was in the air, I decided that I did not have enough space in which to land, so my alternative was to fly to the right over the trees and somehow land on the main airfield. I managed to turn my machine and avoid the trees, and after flying a short distance further I was ready to land.

Alexander Riaboff at the controls of a Farman 4 trainer in 1916. Flight training begin with this durable biplane and then progressed to more advanced aircraft.

Riaboff sits on the forward edge of his Farman trainer after a landing mishap at Gatchina.

This fuselage section was used at Gatchina as an improvised ambulance and fire truck.

I was confronted with a slight problem at this juncture, for I had not as yet been taught how to land! I was probably fifty feet above the ground when I figured that if I kept the Farman level and reduced its speed I would eventually land somehow on the vast practice field at Gatchina. After cutting my speed, I did my very best to keep the Farman horizontal. The aircraft began to settle down all right, but as I approached the ground I noticed that in spite of all my efforts, the tail was going down faster than the nose. I hit the ground; the landing gear was smashed and the lower wing was resting awkwardly on terra firma. Fortunately for me, I was neither scratched nor frightened by the accident, but I was extremely upset.

I climbed out quickly and surveyed the damage. I was mad as a hornet. Another cadet came over to console me. "Don't take your accident so seriously," he said. "Every student has at least one foul-up with his trainer!"

"But, don't you see," I replied, "the government entrusted me with this aircraft and I've wrecked it. I feel awful. How much do you think an airplane like this one costs?"

"I guess about three thousand rubles," he ventured. "Three thousand rubles! Oh my God, and to think I damaged it!" I was horrified at the thought of such costly damage.

My new friend was persistent: "Don't worry so much. The trainer can be repaired," he assured me. I left the scene with a determination to learn where I had gone wrong in my attempt to land for the first time in my life. Later, my instructor carefully explained and demonstrated to me that in order to land a Farman 4 the pilot must first nose the aircraft downward, then cut the motor, keeping the descent at a steep angle until it almost hits the ground, at which time it must be leveled off. If the pilot fails to nose the plane downward first, and instead cuts the motor first, the Farman immediately loses speed—and then it becomes impossible to nose it downward or keep it horizontal. His explanation described fully my mistakes, which had resulted in an accident.

I could hardly wait to tell my family what I had done. By this time I looked upon the incident as being a good joke on me, and I described the event in a humorous style in a letter to my family. To my surprise, they were filled with fear and apprehension for my safety. Flying to them was not a funny business. My accident confirmed all their fears.

After we learned to taxi the Farman 4, our instructor explained the main points of handling it in the air. Lessons began with the instructor seated in the front of the trainer and the student in the rear. As both men were fully exposed to a stream of air, oral communication between them was impossible, so they used a primitive form of sign language. Training flights were of short duration around the airfield and at low altitudes, since

A Deperdussin monoplane lands at the Gatchina airfield.

This French-designed Morane Parasol monoplane is equipped with skis for winter operations at Gatchina. Riaboff made a dozen flights with this aircraft while training at Gatchina. By 1916 the Morane had become obsolete, and many of these aircraft had been reassigned to the various flight schools for training purposes.

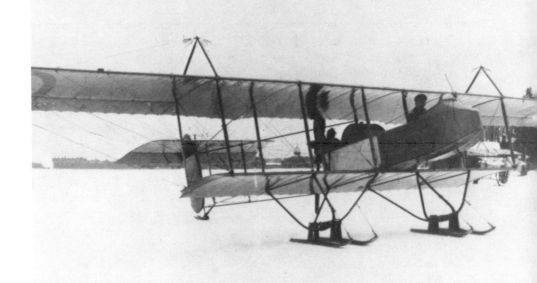

Alexander Riaboff had entered the final phase of his training at Gatchina when he took the controls of this Farman 22 advanced trainer, equipped for winter operations.

A Voisin bomber at Gatchina. This durable French-designed aircraft was used widely by the Imperial Russian Air Force in 1916.

their main purpose was to teach us to take off and land. After several flights to familiarize the student with the art of flying, he was permitted to hold the stick, which could be easily taken over by the instructor at any moment. When the cadet had mastered these takeoffs and landings, he was allowed to execute the shallow turns by tilting the airplane. Finally, he was ready to solo above the airfield at an amazing altitude of two or three hundred feet. To solo was a major milestone for each person, an event that marked one's maturity as a flying cadet at Gatchina.

Some cadets were afraid of heights and therefore tried to fly low, but I attempted to keep a reasonable distance between the airplane and the ground because I wanted as much space as possible for maneuvering. I kept in mind that a crash always involved a collision between the trainer and the ground. I had no difficulty with my solo, but in subsequent flights my motor stopped a few times when I was several hundred feet up, and yet I was able to land successfully.

After mastering the Farman 4, I was transferred to a group flying the Farman 16, a more advanced trainer with a larger, more powerful engine. It had more speed, could glide a long distance, and was easier to fly. In time, I successfully completed my flights with the Farman 16 and was assigned to a group of seasoned cadets flying Farman 21s, which at that time were the most modern trainers at Gatchina. Except for my first unauthorized flight, I completed my entire flight training at Gatchina without incident.

Some students were not as fortunate. The first cadet to die in a crash was a likable chap from a well-to-do family. He was attempting to land a Farman 16 when the trainer fell into a vertical dive for some unknown reason, and he was unable to pull it out. His funeral was the first one our group attended. Another student, a talented, well-educated chap, was also killed while trying to land. Admittedly, he was a poor athlete, uncoordinated, and probably unfit to fly.

There were the usual accidents, seemingly caused by some quirk of fate. A student in our group asked me if I would take him for a flight in a Farman 16, as he had never flown in one. I agreed, but while I was waiting for my turn to fly, he changed his mind and went up with another student. Ten minutes later, as they were approaching the field to land, their trainer suddenly nosed down sharply. I saw a body hurtling out of the passenger section, falling with hands outstretched. The trainer immediately leveled off, at which time the student pilot was informed that his passenger had fallen out. It was winter and the wind had blown snow into a drift as high as a one-story building. My friend had fallen into this large snowdrift and was quickly extricated, but he was found dead, with practically all of his bones broken. These fatal mishaps came suddenly and they reminded us of

Полошка Шт.-к. Файбишевича

Some aircraft accidents at Gatchina ended in tragedy. Staff-Captain Faibishevich was killed while testing a Farman trainer that had been refitted with a new 70-horsepower Gnome engine.

A funeral procession in Gatchina for a cadet killed in an abortive attempt to land a Farman 16 trainer.

Сгорбшая угла 13/5 16.

The fabric wing covering of this Farman 4 was burned off in a fire.

Training at Gatchina was punctuated with flying mishaps. An embarrassed cadet, a Lieutenant Filimonov, stands at center, holding his flying helmet, after he lost control of his aircraft and fell into a ground loop while attempting to land on an overcast October day in 1916. His instructor, Lieutenant Chepurin (center, to Filimonov's left), looks on. Riaboff stands next to Filimonov.

the high risks associated with our training.

I soon came to realize that the daring, reckless fellows, the cadets who took unnecessary chances such as making sharp turns too close to the ground, did not last long. In fact, one of our experienced instructors, while flying a Farman 4, was killed himself while demonstrating sharp turns too close to the ground. If he had been 3,000 feet above the ground he would have had space to dive downward, attaining enough speed to make the trainer maneuverable.

There were the timid, overly cautious students, quite often poor athletes, who were actually afraid to fly. They also failed to last. It was my observation that the fairly good athlete, a person able to make quick decisions and execute them, had the best chance to survive.

While on field duty one time I witnessed a miracle. I noticed a Farman 21 about a half mile away and approximately 2,000 feet above the town's park, descending at a normal angle, evidently preparing to land on the field. Suddenly, for no apparent reason, the aircraft nosed down and proceeded to fall vertically. I expected it to level off, but alas, it did not and soon disappeared in the tall trees nearby. I called an ambulance immediately and informed the staff about the accident, convinced that the pilot had been injured or killed.

I abandoned my post and ran to the park in the vicinity of where the Farman should have fallen. There I encountered a middle-aged officer coming toward me. His face was bruised in several places and he appeared to be walking in a stupor, with a wandering look in his eyes and carrying a pilot's helmet in his hand.

"Is that you sir?" was the only question I could muster in my shocked state of mind. He simply nodded affirmatively, unable to speak at that moment. "Just proceed along the road," I told him, "they are expecting you." I stood there transfixed, unable to believe what I had just seen. I had witnessed a 2,000-foot, vertical fall of an airplane right to the ground, and the pilot had not only survived, he was walking! How could such a thing have happened?

I then ran to the scene of the accident to observe it carefully. There stood an exceptionally large, very tall pine tree next to an elevated road about twenty-five feet above ground level, with a bank of dirt extending from the ground to the road at a forty-five-degree angle. The aircraft in its downward plunge had hit the tree with its left wings near the top, shearing the branches off with its fall. The wings were completely destroyed. Pieces of the tail surfaces and the wings remained in the tree.

The striking of the branches by the left wings and by the tail surfaces had retarded the machine's fall, particularly on the left side, so that the aircraft had fallen with its right wings directed down at a certain angle and slightly forward. The right wings had struck the top of the road and had been sheared off. This peculiar sequence of events left the front part of the fuselage with the pilot intact. At the moment of impact the downward motion of the fuselage roughly paralleled the incline of the road's bank. Consequently, the fuselage slid along the bank, leaving a deep, but gradually diminishing, burrow until reaching ground level. Thereupon, the pilot stepped out of his truncated aircraft, having completed his unusual landing!

Our flight training progressed slowly because the Gatchina school needed more and better training aircraft to replace the obsolete French models we used. We needed better engines and more spare parts, but our primary need was better management of the school and a single, coordinated, positive training program. The flying school's staff failed to recognize the fact that some men were not pilot material. They tried to convert every student into a good flier, which was an impossibility and resulted in an unnecessary waste of time, materiel, and lives.

Life at Gatchina was quite monotonous until a new rule was introduced that gave us the choice of living at school or in private homes. My former classmate at the Alexander Commercial School in Moscow, Vanya Novozhilov, and I agreed to live together. We located a lower flat for rent in a building close to the school. It had two bathrooms, a living room, and a kitchen. Now we realized we would have much more freedom to circulate among the local inhabitants and to enjoy a social life for the first time. I must admit that we young pilots were much sought after as companions.

In time, our graduation from Gatchina Military Flying School approached. Our school's administrators designed a series of examinations leading to promotion as ensign. Aviation at that time was officially with the Engineering Corps, so it was suggested that we be given a special examination at the School of Army Engineers, as though we were a graduating class. Because we were students of aviation, they promised to change the examination subjects slightly by substituting our disciplines for military engineering.

After a month's preparation, we went to the School of Army Engineers, referred to as "Engineering Castle," formerly one of the elegant Romanov residences, to take the examination. Every facet of the test went well except for the part on how to command a platoon of soldiers. Although we knew how to march and take commands, we had had no practice in giving orders.

The first rule of a commander was to declare loudly: "Platoon, obey my orders!" Only then are commands obeyed. Our examiners took us to a large field on which stood a number of old, three-story brick buildings occupied by a regiment of well-trained infantry guardsmen. Also, there were numerous high piles of firewood stacked in various places on the

First Solo

[Postcard addressed to Vassiliy Ivanovich Riaboff, Krasnoselskaya Street, Building 66/16, Apartment 10, Moscow]

August 18, 1916

Dear Father!

Today I soloed for the first time! All alone for the first time in the air! I flew at an altitude of seventy feet. I flew this airplane with calm and assurance.

On the face of this postcard is a photograph of a Farman 4 trainer that we use at Gatchina. It is nicknamed the "Fourth" or the "trainer." I have flown the Farman pictured on this postcard, number 26, several times with our instructors. You will notice the chassis is damaged on the aircraft.

And now until I see you.

A. Riaboff

A group of officers at the Gatchina Military Flying School. Some officers in the Imperial Russian Air Force with combat experience were rotated back to Gatchina to serve as instructors.

Cadets at leisure in the dormitory of the Gatchina Military Flying School. This photograph was taken by Alexander Riaboff.

field. A platoon of infantry guardsmen and several engineering corps officers, our examiners, were awaiting us. A student was called to take command and move the platoon forward.

"Platoon, forward march!" the student commanded in a loud, clear voice. Not a man stirred. Obviously perplexed, the student looked at his examiner, who solicitously suggested, "Probably they didn't hear you. Make your command louder." Suddenly, the light dawned and my colleague recalled the magic words, "Platoon, obey my orders!" and they did—precisely and beautifully.

An army examiner then told another student to move the platoon to the right, left, forward, and again to the right. When the platoon marched straight to the wall of stacked wood, the officer ordered: "Stop the platoon!" Being mentally unprepared for the order, the student remained speechless while, to his horror, the soldiers continued to march, climb over one stack of firewood, down the other side, and onto the second stack just like an army of peasants.

"You see what you have done?" the officer shouted. "The platoon has broken formation and is climbing over the stacks. Do something! Restore order!" In spite of his rapid-fire commands, I noticed a look of amusement on the officer's face as the "army of ants" continued to march forward, relentlessly, until absolute confusion reigned.

"Stop them," he demanded. By this time the student had completely lost control of himself and was unable to utter a word. Seeing his state of helplessness, his examiner took over: "Platoon, obey my orders! Platoon halt!" The soldiers halted wherever the command caught them. They restored formation quickly and that particular "game" ended.

During the grim winter of 1916–1917, I completed all the requirements, having soloed in August, to become a military pilot. Those of us who passed these examinations at Gatchina were given the rank of ensign and, in accordance with the existing law, Nicholas II personally signed the order of promotion. Soon afterward, in March 1917, the emperor abdicated. We were the final group promoted to first officer's rank by the last Romanov Emperor of Russia.

Our promotion called for a celebration, so twelve of us, with newly acquired girl friends, gathered together in the flat my friend Novozhilov and I rented. It was the first party to be given by the cadets in our group. Everyone was in a jubilant mood, especially after a few rounds of straight vodka. Within a short time the party became hilariously unrestrained; the noise was deafening as we amused ourselves by staging a contest after midnight to see which one of us could kick the upper plank of the door frame. Each attempt to reach it was accompanied by cheering or booing, and the party reached its zenith when Vanya finally kicked it and won the contest. The tumultuous uproar that followed brought a sharp, determined knock at the door.

There was dead silence. Everyone froze and stood at attention. Who could it possibly be? Someone then gingerly opened the door and before our horrified eyes stood a general in full regalia. "Attention!" a voice from our group commanded.

"Gentlemen," the general began, "I want you to know that my wife, two children, and I live in the flat above you. We were aware of the fact that two of you had moved into this flat, and up to now the occupants have lived quietly and we have had no complaints to make. I realize you are having a party and you wish to enjoy yourselves fully, but could you please not be so boisterous! My wife and children are hysterical. The house has been shaken to the extent that dishes falling from the shelves have been broken! Kindly cease whatever you are doing—quiet down and please spare this house from complete destruction." He then departed.

For a moment no one spoke, then one fellow volunteered, "At ease friends. That chap is not a real general. He's some kind of civilian general." Although his observation reduced some of our tension, our unforgettable celebration had ended abruptly.

Vanya went to Moscow for a few days to celebrate Easter with his parents. While he was away the two of us received an invitation to attend a private party and he was asked to join us. To everyone's surprise, he didn't appear until midnight.

"Vanya, what took you so long?" I asked, "Where have you been?"

"I just returned from Moscow. The train was packed to overflowing the entire trip and I can't stand up another minute!"

"Aha, what you need is a big shot of vodka. Here, drink it down," one man advised Vanya, thrusting a tumbler of the liquid fire into his hand. "You'll be as fresh as a cucumber and ready to go another twenty-four hours!"

Vanya swallowed the entire amount and soon thereafter slumped down on a couch, passed out cold and dead to the world. All attempts to awaken him failed, and when the party ended we were faced with the problem of what to do with him. Public transportation was unavailable so our only choice was to carry my friend to our flat. Revolutionary times required the streets to be patrolled nightly, and none of us wished to become involved with the authorities, as every aspect of life was precarious.

Six of us transported Vanya, four men at a time, each holding onto a leg or arm. Suddenly, one of the body bearers, who had a powerful bass voice, burst into song. Being half drunk, he intentionally sang old Russian patriotic songs—a dangerous thing to do. Eventually, a patrolman

The Russian winter proved to be no insurmountable barrier to flight training at Gatchina. Here a pilot, perhaps an instructor, prepares to take off from the cold and wind-swept Gatchina airfield.

Officers and cadets gather for a Russian Orthodox divine liturgy on December 6, 1916.

approached us to take a look at the presumably lifeless Vanya, whom we had deposited on the ground. "Hey, what's going on here? Oh ho, you have a body there. What are you trying to do with it? Dispose of it, eh?"

When it dawned on us that the patrolman assumed we were carrying a corpse we laughed uproariously. Upon regaining our composure, a very touching, poetic speech was delivered by our basso, which I'll attempt to paraphrase: "My dear, noble man, you are under a delusion. This body is not dead. Touch him. He is warm—hence alive and happily reposing in Nirvana. You wonder how come? Well, you see he drank vodka. Vodka, my friend. Do you understand what that means? Have you seen a man lying on the street drunk from vodka lately? No, of course not. Just touch him and be proud to tell your friends that on this memorable day you touched the happiest man in the whole world!"

His speech had genuine meaning to the vodka-drinking Russians, since the sale of vodka had been prohibited for the war's duration. There was a trace of envy in the patrolman's eyes as he gazed upon the prostrate body lying at his feet. Vanya was delivered safely to our flat and deposited on his bed, where he slept off his inebriation and, except for a large hangover the following day, soon became his normal self.

Our celebration, as events proved, was a brief interlude before we faced the grim consequences of Russian defeat on the battlefield and the growing anarchy of the Russian Revolution.

A caricature in snow! A Gatchina cadet stands between two snow sculptures of Emperor Franz Joseph (left) and Kaiser Wilhelm II, Russia's wartime enemies.

A Farman 4 lands at the snow-covered airfield at Gatchina.

The Farman 4 primary trainer was equipped with skis for flight training during the winter months of 1916–1917. Alexander Riaboff is at the controls for a training flight.

3. Into the Revolution

At the time of Alexander Riaboff's military commission as an ensign in the Imperial Russian Air Force in 1917, Russia was still under the Romanov dynasty. By March 7, a massive popular revolt in the capital of Petrograd forced Nicholas II from his throne. Workers' strikes, street demonstrations, and mutinies within the armed forces had brought the 300-year-old Romanov dynasty to an end.

A Provisional Government assumed official power, having been created by a committee in the Duma, or parliament. At the same time, revolutionaries set up the Petrograd Soviet of Workers and Soldiers Deputies to challenge the authority of the Provisional Government. Soviets, or councils, of workers and soldiers spontaneously appeared in most of Russia's major cities.

The Provisional Government, composed of more moderate political elements interested in building a constitutional democracy, lasted for a mere eight months. This regime, under Alexander Kerensky, was overthrown by the Bolsheviks on November 7, 1917.

Russian political life was shattered in 1917, giving way to social upheaval in the cities and the countryside, the collapse of the Russian war effort against Germany, and a brutal struggle for power between Russia's various political groups. Being close to Petrograd, Riaboff saw the epic Revolution firsthand and, as a young officer, he faced its many dangers.

■   ■   ■   ■

M Y FLIGHT TRAINING AT GATCHINA spanned the dramatic winter of 1916–1917, the grim period of military defeat and growing popular unrest that led to the Revolution of March 1917. Russia had been ill-prepared for the demands made upon her by World War I. We entered the war unprepared. Our war effort could not overcome the nation's technological backwardness, lack of industry and means of transportation, and the government's failure to supply munitions and other war necessities for

Cadets from Gatchina made numerous flights over Petrograd. Alexander Riaboff took this picture of the city, on the approach to the Gulf of Finland, in 1916.

her troops. As the war continued, our supplies became exhausted and replacements were insufficient. Logistically, the situation became impossible.

Many of our infantry went to the front without rifles. They were given orders to pick up the guns abandoned by the dead and wounded. Despite the bravery of the individual Russian soldier and early victories, the Russian Army found itself in retreat in 1915. Lack of supplies was attributed to the short-sightedness of the government and the bungling of the army's general staff. When the Duma began to stress the government's derelict conduct of the war, a permanent rift arose between the two. The Duma first dealt with the problem of supplying the army by establishing a great number of committees throughout the country in order to organize and oversee the production of military necessities. Soon, the Duma became the nucleus of a huge bureaucracy. The crisis, however, persisted, and popular unrest grew.

Nicholas II, hoping to raise his fading prestige with his people, the army, and Russia's allies, named himself supreme commander of the armed forces in 1915. As commander, he moved to the front and gradually became politically insulated from the rest of the country. The Duma had no illusions regarding the emperor's capabilities or his capacity for leadership as head of the army or head of the government. With the emperor occupied at the front, the rumor circulated in Petrograd that the Empress Alexandra had become the real power in Russia. She in turn had fallen under the influence of Rasputin, an uneducated peasant "prophet" from Siberia who was reputed to have exhibited a remarkable power to stop the bleeding suffered by Tsarevich Alexis, a hemophiliac.

The name Rasputin in translation means "dissolute, lewd, profligate," and these qualities applied to the man perfectly. His reputation was that of a vicious, obscene debaucher who readily sold his palace influence for a good party with young women. In fact, his power even extended to the emperor himself through the empress, who influenced important state decisions according to the advice of Rasputin.

On December 30, 1916, newspapers published the story of Rasputin's death at the hands of Prince Felix Yusupov, who was related to the imperial family by marriage; Grand Duke Dimitriy Pavlovich, a nephew of the tsar; and V. M. Purishkevich, a member of the Duma. Unusual details of his murder revealed that Prince Yusupov had invited Rasputin to a party at his palace in Petrograd. During dinner, French pastry and a goblet of wine, both heavily laced with poison, were served to Rasputin. While the conspirators fed Rasputin the poisoned fare, a doctor stood in the wings observing the grim drama.

To everyone's surprise, Rasputin consumed the poisoned treats with no visible effect. In nervous desperation, Prince Yusupov then fired several shots into Rasputin's body. Even this tactic seemed to fail. Riddled with bullets, the evil one crawled on hands and knees through a passageway of the palace into the snow-covered courtyard where he finally collapsed and was declared dead by the attending physician.

Yusupov and his friends then took the body to the Neva River, pushed it through a hole in the ice, then shoved it under the ice to make sure it would remain submerged. Several days later the body was discovered, and an autopsy was performed to determine the cause of death. Surprisingly, it revealed the fact that Rasputin had died from drowning and not from bullet wounds or poison. This grisly episode gave the winter of 1916–1917 a macabre touch.

"If I die, Russia will perish!" Rasputin had once prophesied. Indeed, the political chaos that followed his demise, it seemed, made his words an ill omen! Following his death, a general sense of relief prevailed, and his murderers became heroes to many people. Yusupov and his accomplices were neither arrested nor tried in a court of law. The government merely banished him from Petrograd, an order given by Emperor Nicholas II. Time for the old regime was short.

When Nicholas II departed again for the front in February 1917, demonstrations and disorders broke out in Petrograd over the lack of food. Soon, the city was overrun by revolutionary mobs roaming through the streets. Policemen and army officers were killed, a jail was forced open, homes were set afire, and some reserve battalions actually joined the revolutionary crowds. Anarchy had taken over completely. On March 15, 1917, Nicholas II finally abdicated and the Duma immediately established a Provisional Government that proclaimed its intention to continue the war with Germany to its victorious end!

Unfortunately, the Duma's leaders lacked the qualities necessary to lead Russia out of this morass of anarchy. They tried to govern according to the lofty principles of law and democracy, whereas Russia needed a determined, willful statesman at her helm. Imperial Russia, ruled by the Romanovs since the seventeenth century, more than three centuries, now ceased to exist. Holy Mother Russia no longer existed.

Professional revolutionaries, especially Bolsheviks, immediately organized and plotted to take power. They violently opposed the Provisional Government and organized "soviets" to represent workers and soldiers, and agitated for more radical changes. Under Vladimir Lenin's strong leadership, the Bolsheviks won their struggle against the Provisional Government and declared themselves the victors on November 7, 1917.

My aviation school at Gatchina somehow continued to operate normally at the beginning of 1917. Gatchina was a small town without any industry of importance. Most of its middle class residents worked in Petrograd. Comparatively few workers lived there. Consequently, demonstrations and disorders were nonexistent in Gatchina, although disturbing reports of chaotic conditions in Petrograd inevitably reached us.

Strikes, prompted by severe food shortages, paralyzed the capital. These strikes and demonstrations involved hundreds of thousands of workers employed in the industrial plants of Petrograd. Many of the demonstrations took a violent turn, with street fighting between strikers and the police. Every day shops were looted and burned. Policemen and military officers were murdered. Entire regiments joined the strikers, and battalions sent to quell the violence refused to shoot into the crowds. An officer risked his life whenever he appeared on the streets in Petrograd. These events prompted uneasiness, even fear, in Gatchina.

Despite many dire warnings, a group of us visited Petrograd at the time of the March Revolution. The mood of the city had changed. On the Nevsky Prospect, the capital's most important street, unusually large crowds of civilians, with a generous admixture of soldiers and sailors, milled about. Each passing face seemed to register excitement. Heated political arguments filled the air and sometimes exploded into actual fighting. Those who believed in the Revolution marched and sang, assured of its eventual victory.

Watching the scene intently were groups of spectators made up of some officers, intelligentsia, and the middle class, who stood apart and aloof from the activities. They appeared to be apprehensive, confused, and worried about the unpredictability of their futures. My own feelings were that the Revolution was untimely and that the continuation of the war would be the first order of business. In my pessimistic opinion, the call to win the war and to have a successful, so-called bloodless revolution simultaneously appeared impossible.

While exploring some of the streets adjoining the Nevsky Prospect, I encountered a large number of soldiers milling about a large gate leading to a military compound. I realized that I was approaching the barracks of a reserve regiment where fresh recruits were receiving military training. The general atmosphere surrounding the crowds of soldiers gave me the impression that the Revolution had taken entire precedence over the war.

Suddenly, I became aware that I, as an officer, was being observed, and some of the men were showing signs of becoming agitated in anticipation of what they feared I might do. Across the street, a small group of soldiers stood idly about. Momentarily, I hesitated as to my next move. If I retreat from the crowd, I reasoned, it might indicate that I am afraid, and this could be the catalyst to start something! I decided to walk nonchalantly straight toward the crowd, recalling all the while that officers had been

An aerial view of Petrograd (formerly St. Petersburg), taken in 1916.

molested, beaten, and killed for no reason except for the fact that they were army officers.

When I reached the crowd, at least one hundred pairs of hostile eyes focused on me, and then, to my great relief, the mob made a passageway for me. I saw many faces but remember only one in that crowd. It was that of a young peasant with widely set, innocent eyes and an open expression—a harmless type by himself, but, nevertheless, ever ready to follow the mob. We exchanged smiles as I passed through another encounter, unscathed. No one had made a remark, a threat, or even challenged me. I can only deduce that the combination of my naiveté plus common-sense in the face of danger had saved my neck.

It was shortly after returning to Gatchina that I completed my last flying examination. The test included a flight to and over Petrograd at about an altitude of 7,000 feet. This particular flight would remain vivid in my mind. The sun had nearly set, and as I flew over the great city, often referred to as the Venice of the north, I witnessed a spectacular sight. The Gulf of Finland appeared to be a sea of blood as it reflected the sunset held within a framework of ebony shorelines.

Could this be the apocalypse of Russia's future? Does it indicate that a sea of blood will be shed on the altar of revolution? My poor, unfortunate Motherland, I thought. You have been tested many times before, and now again many millions of your sons and daughters must suffer and perish!

Richly endowed Russia had been sought after by the European powers in previous wars and I knew there were countries that would divide her into colonies and rejoice to see her perish as a nation. Time alone would tell if my country were to survive. I flew back to Gatchina, cut my motor, and glided into a landing. True enough, I had passed the flight examination, but I had also experienced a disturbing portent of the future.

While I was at Gatchina as a commissioned officer, my superiors assigned me to the post of Officer on Duty in the auxiliary aviation battalion, which was located next to our flying school. So far as I knew, no officer from the school had ever been assigned this duty before, but at the beginning of the Revolution the impossible became the possible and vice-versa.

These revolutionary days were dramatic, always unpredictable, and dangerous. Old laws were not abrogated, they were simply disregarded and were not replaced with new ones. Therefore, anyone with any power began to act as he saw fit or by his own idea of patriotism. Every facet of Russian social existence was chaotic. One well-known journalist was overwhelmed by these events. In desperation he published an article suggesting that the revolutionary government should pass laws, laws of any kind, even stupid ones, as the country was rapidly heading toward anarchy. It was an "every man for himself" situation.

When I assumed my new post, I hadn't the slightest idea of the duties or authority of an Officer on Duty. I had no experience at command. Fortunately for me, I had an experienced, intelligent sergeant as my assistant. I went on duty at noon, and shortly thereafter things began to happen. My sergeant reported that the supervisor of the former Imperial Gatchina Palace on the tsar's hunting estate adjoining the city had come to see me regarding a very important matter. I instructed the sergeant to admit the man. He proved to be a very intelligent, cultured, elderly gentleman who was entrusted with the care and maintenance of the tsar's palace.

"I have come to complain," he said, "about a group of soldiers who have occupied my office, are interfering with my management of the palace, and, worst of all, refuse to let me have oil with which to heat the building. Without heat all of the rare, valuable antiques will be permanently damaged by the cold and dampness. Furthermore, sir, the soldiers have helped themselves to a very valuable collection of cigarettes and are smoking them. Please, sir, order the soldiers not to interfere with my management of the palace. Ask them to release the oil supply and tell them to refrain from using or taking anything in the premises."

I decided that his request was reasonable and should be complied with immediately. Accordingly, I prepared, signed, and issued an order covering all of his requests. "Thank you, sir. Your act is the first sign that sanity may be returning to Mother Russia and the Revolution!" We shook hands and concluded our meeting with mutual satisfaction.

My next audience was with a baker who wanted a permit to continue his bakery business. "For some unknown reason a certain group of people will not allow me to bake," was his complaint. I issued a permit to him to operate his bakery. Three women followed with a complaint about a group of soldiers who were trying to evict them from their flat with the aim of occupying it themselves. I issued a written order specifically stating that "the women are not to be bothered."

Toward evening the sergeant suggested that we conduct an inspection tour of the state vodka store. Earlier, a platoon of soldiers had been stationed there to guard against possible break-ins, thievery, or destruction of the precious liquid. Not being at all certain of what I would find at the store, I asked the sergeant to provide me with a platoon of soldiers for the expedition. We marched through the town, arrived at the store, and found it intact but with its contents somewhat depleted. A happier guarding platoon I have yet to witness!

I relieved them of their duty, sent them "marching" back to battalion headquarters, and left half of my platoon to guard the place for the next twenty-four hours. During the night the sergeant and I returned to check up on my charges and found everything normal. While there, however, I chanced to come upon a Moscow acquaintance, a former fellow student who served as a soldier in the battalion. Although I had not known him well, he confided to me that he had become a revolutionary and was very proud of the fact that he was spreading revolutionary propaganda among the other soldiers. To emphasize his commitment to the movement, he produced a large German revolver and gave me to understand that he would not hesitate to use it against counterrevolutionary elements, including officers.

Could he have been threatening me? I wondered. My question was not answered, as our paths never crossed again. Around midnight the sergeant reported that a bunch of thieves and pickpockets had been arrested at the Gatchina railroad station and had been taken to our post.

"What shall we do with them?" I was asked. "Bring them to me and I'll question them," I replied.

Twenty boys ranging in age from twelve to fifteen years were brought in. I was told that they had been riding trains between Petrograd and

Gatchina and stealing whatever they could from passengers. At first they would not talk, but as I continued to question them I learned that most were homeless street urchins living on their own in Petrograd. Only a few had families so I decided to return them to police headquarters in the city, since there were no facilities in Gatchina to handle them.

"Sergeant," I ordered, "select a few soldiers to accompany and guard the boys until they are safely delivered to police headquarters in Petrograd." I felt at the time that I had made a wise decision. My new experience as Officer on Duty was exciting and exhausting. Revolutionary chaos produced nearly insurmountable problems that had to be solved quickly. Yet I was to learn very soon that remedial solutions were short-lived, in some cases, regardless of the fact that I considered each problem thoughtfully and as reasonably as my experience permitted.

During my twenty-four-hour tour of duty I was confronted with a steady flow of requests and complaints, but I have described only those episodes that impressed me the most. The following day I was relieved of my assignment. Two weeks had passed when I chanced to meet my former sergeant and we recalled the events of the last day I had the duty. I asked him if he knew what had become of the arrested boys.

"Oh, those boys," he said, and his face suddenly registered an expression of dejection. "Well, it seems they were on their way to Petrograd when the convoy decided that it was a waste of time to take the young criminals to the city so they were taken off the train and shot!" He shook his head as if to negate the horrible episode, and he was visibly disturbed when he saw my absolute distress.

"My God," I moaned, "it was not my intention to send those poor boys their death. How could I have foreseen such a tragic end for them!" I was distraught beyond words and felt completely helpless when I realized I could not rectify my dreadful mistake. Was this an example of what happens during a so-called "bloodless revolution"? I was soon to learn that blood was beginning to flow in ever-increasing streams. Life had become a cheap commodity. People died on the streets day and night—victims of the "bloodless revolution."

While I contemplated the tragic deaths of the Petrograd urchins, the thought struck me that no one would have cared for those homeless, neglected boys. I was informed later that no one had bothered to investigate the circumstances surrounding their deaths.

My next assignment was to go with a platoon to a large railroad station thirty miles southwest of Gatchina to restore order. Whoever telegraphed the request for help had neglected to explain why it was needed, so my commanding officer instructed me to appraise the situation and act accordingly. I arrived at my destination to find the station quiet and virtually empty. When I located the stationmaster, I asked him if he knew who had called for help and why. I could see that he was unfavorably impressed with my youthful appearance, and his face showed disappointment when I told him I was in charge of the platoon.

"I'm the one who telegraphed for help all right," he admitted. "The local railroad workers are planning to have a meeting at the station today and I'm afraid for my life!" he explained excitedly. Shortly thereafter, workers began to file into the station, followed by a small group of organizers. "Tell me, please, why are the people gathering here?" I asked a man who appeared to be a leader. "We workers decided to have a meeting at the station because of its convenient location. We want to discuss many of our problems—problems not connected with the railroad station," he explained.

I told the stationmaster the substance of my conversation with the organizer and said, "I see no immediate trouble for you and I'm returning to Gatchina. Should any problems arise, just telegraph us and we'll be back," and we boarded the first train returning to Gatchina. During the trip, one of my soldiers came to me and reported, "Sir, there's a "counterrevolution" riding in the next car!" I restrained a laugh as I accompanied him to the next car to witness my first "counterrevolution." The subject of his complaint proved to be an elderly, well-dressed lady of the former nobility. She was surrounded by a hostile crowd of peasants and workers.

"Pardon me, Madam, I am Ensign Riaboff. May I be of any assistance to you?"

"Oh, thank you, Ensign," she replied nervously. "I am returning to Petrograd from my estate that was recently taken over by peasants in the village. Why—why have they done this dreadful thing to me? I have never harmed anyone. My pleading with them has been to no avail. They drove me off my land!" She spoke bitterly, trembling with anguish. Unfriendly faces hovered close to her and she obviously feared for her life.

"Don't worry, Madam, I'll see you safely to Gatchina."

I glanced about me, looking directly into those faces that appeared to be the most threatening, and declared, "No one will harm you while I'm here." Turning to the soldier I said, "You have made a mistake. This lady is old and ill. She can't harm the Revolution's cause in any way." We arrived in Gatchina without incident.

In retrospect, I am amazed that in those precarious times someone as naive as I had been placed in positions requiring sophisticated decisions. Perhaps, because of my innocence and my inability to grasp the seriousness of many situations and the grim consequences that might arise from them, I was spared many difficulties. In other words, "Fools rush in where angels fear to tread."

4. Assignment in Odessa

Having completed his training at Gatchina in the spring of 1917, Alexander Riaboff attended an advanced flight-training program for fighter pilots at Lustdorf, near the strategic port city of Odessa. This was an excellent assignment and no doubt reflected the confidence of his instructors at Gatchina.

When Riaboff reached Odessa, Russia was still at war. The Provisional Government was anxious to prosecute the war to a successful conclusion. By contrast, the country was in social and political turmoil, for the most part disengaged from the war, and overwhelmed by the growing anarchy. A. F. Kerensky, Minister of War, launched an offensive in July 1917 against the Germans and Austrians, hoping a quick victory would bolster the Provisional Government. A German counteroffensive crushed Kerensky's hopes and simultaneously strengthened the hand of the Bolsheviks, who had openly opposed the war.

Riaboff was at Odessa when the Bolsheviks assumed power in November 1917. He would remain at the flight school through February 1918. During this dramatic period he observed the rapid growth of revolutionary sentiment and eventual Bolshevik control of Odessa, which had been a center of revolutionary activity going back to the 1905 Revolution, when sailors mutinied on the battleship *Potemkin*. This same warship, ironically, would play a part in Riaboff's difficult days at Odessa in 1917.

During this same period, the Germans invaded the Ukraine, at the time a region gripped by anarchy, with nationalists, Bolsheviks, and assorted revolutionaries contending for power. The Treaty of Brest-Litovsk, which ended hostilities between the Germans and the Bolsheviks, was signed on March 3, 1918, shortly after Riaboff had left Odessa for Moscow. The experiences of Riaboff in Odessa reflect the revolutionary chaos of 1917–1918.

■　　■　　■　　■

A Morane trainer in a landing mishap at Odessa.

IT WAS NOW TIME FOR THE PILOTS at Gatchina to be reassigned elsewhere. Some were sent to the Baku Maritime School of Aviation, some to the front, and some to the "School of High Pilotage" at Odessa to be trained to fly the latest fighters. The latter assignment was highly coveted and several of us, including myself, were sent to Odessa for advanced flight training.

While en route to Odessa, I had time to stop off in Moscow to visit my family. As the cab drew up in front of my home I had the feeling that I had become an entirely different person from my former self. Now I was an officer pilot, a much older man who had witnessed death, revolution, and drastic changes, and who had begun to sense the many complexities and uncertainties in life.

My mother, father, and three sisters greeted me with open arms and expressed great joy in finding me alive and well. My good friend Sergei Popov and I met for a good heart-to-heart talk about our respective activities. He remained adamant about continuing his studies at Moscow's Institute of Communications.

"My goal in life is to complete my engineering studies, but I must admit that the course of events may make it more and more difficult for me to stay away from the battlefront. But remember this, old friend," Sergei emphatically declared as we parted, "I'll continue to do my damndest to

Valerian Przhegodsky.

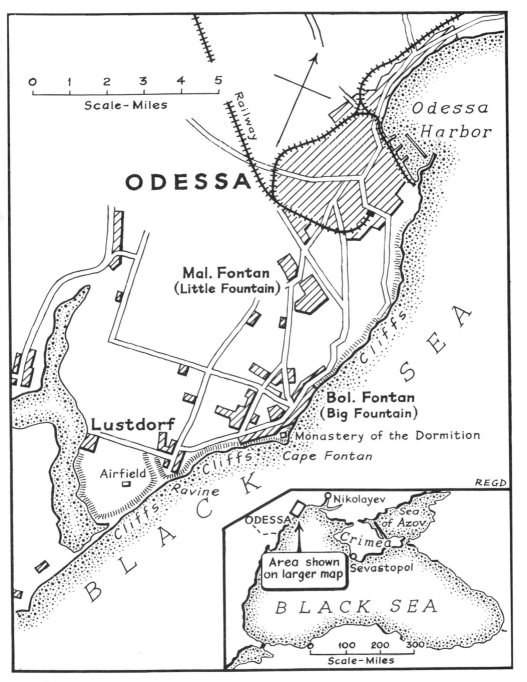

Odessa

Alexander Riaboff was posted to the Odessa Advanced Fighter School in 1917, shortly after earning his commission as an ensign. The school was located at Lustdorf, on the Black Sea coast near the port city of Odessa.

keep out of the mess!" I thought to myself, I wonder if our paths will ever cross again?

The journey by train to Odessa was surprisingly uneventful, considering the revolutionary times. Odessa, a major port on the Black Sea, had a polyglot population of Ukrainians, Russians, Greeks, Jews, Caucasians, and Rumanians. The natives lovingly referred to their city as "Odessa-Mama," perhaps because it was a warm, beautiful place, and life there was much easier than in the fiercely cold climates of Moscow and Petrograd.

Upon my arrival, I sensed immediately that Odessa was a more revolutionary center than the northern cities. Soldiers and sailors had joined the local soviets out of a general revolutionary fervor; they appeared to understand little of the ideology of the revolutionary movement. Who actually governed the city was unknown to us, but, oddly enough, life remained on a bearable course. Our new flight school was located about twenty miles southwest of Odessa near a small German settlement named Lustdorf (meaning Happy Village), located on the Black Sea coast.

I became curious about the existence of a German colony so deep within Russian territory. Investigating the establishment of Lustdorf and other German settlements, especially those along the Volga River, I discovered that in the eighteenth century, during the reign of Catherine the Great, Prussia agreed to exchange thousands of Russians for an equal number of Germans for the purpose of settling in each other's countries. The idea behind the plan was, no doubt, to integrate the various nationalities in order to reduce the probability of war. By the twentieth century, the German colonies had survived intact, preserving their religion, customs, and eighteenth-century German language, and had remained aloof from the Russians. Such was not the case with the Russians sent to Germany, for they had been completely absorbed by the Germans!

We were assigned living quarters in a long, one-story building located on the outskirts of the airfield and just a few minutes' walk to an excellent beach, where we swam as often as possible. My roommate, Valerian Przhegodsky, a Polish chap, and I enjoyed ourselves to the fullest in our newly found paradise.

Military discipline, we soon discovered, was nearly nonexistent in this school. Captain Chekhutov, the commanding officer, along with his assistants and instructors, treated us as members of a large, happy family as if we were on a long, pleasant vacation with flying as an additional enjoyment. The locals and vacationers considered us heroes and, to our pleasure, treated us accordingly. Practically all my life I had worked hard and now, for the first time, I was experiencing undreamed of bliss, and I vowed I would take full advantage of the situation. Our school was relatively small, with about fifty students and three instructors in acrobatic

The Odessa Advanced Fighter Training School. Riaboff was assigned to this rugged wind-swept airfield in the summer of 1917. The windmill at right served as a wind soc. At the upper right is the Black Sea. The small village of Lustdorf, a German settlement, is situated beyond the airstrip and fronts on the Black Sea.

The southern approach to the airfield at Odessa. Above the steep bluff overlooking the Black Sea are some of the aircraft hangars.

The administration building at Odessa was once a private mansion. The spacious structure also served as a dormitory for some of the cadets.

Lieutenant Turenko takes off from the Odessa racetrack in his Nieuport 10 on August 8, 1917. His flight was part of an air show given by the Odessa school for the local populace.

flying. It gave the appearance of a temporary establishment having neither permanent nor temporary buildings. Headquarters and small repair shops were housed in private homes that had been taken over by the school.

The airfield was formerly a pasture, unfenced and bordered by seashore on the south and with a dusty country road on the north. We found out in short order that our airfield resembled a huge washboard, with low ridges running at right angles to our usual path of takeoffs and landings. While landing, an aircraft was often bounced upward by the first ridge it contacted; thus, great skill was required on the part of the pilot to keep it from overturning or ending nose downward. I wondered why the airfield had not been leveled and put into proper condition. Undoubtedly, there was neither the machinery nor the labor to do the job.

At the north end of the airfield were telephone lines. The poles and wires proved to be nuisances, and serious hazards, too. Some students, at the beginning of a flight, amused themselves by staying close to the ground while heading toward the wires; then, by raising the airplane's nose sharply, they would leap over the wires, barely clearing them. If executed properly, this was a simple trick and quite safe, but some pilots started their uplifts a fraction of a second too late, hit the wires with their landing gears, thereby snapping them and forcing their aircraft into nose dives. Their airplanes were usually wrecked and the pilots suffered serious injuries. Students who accomplished this trick boasted about their ability to control their planes as well as their nerves. For some unexplainable reason, no order was given prohibiting this daredevil practice.

Flying hours were indefinite and our instructors were neither punctual nor strict, an expression of the general chaos and anarchy that prevailed, and leisure seemed to be the order of the day. Our flight instructors were first-class acrobatic artists. Lieutenant Turenko was the best, and performed his acrobatics with the precision and grace of an aerial ballet performer. Tall, handsome Ensign Chepurin handled his trainer almost as well as Turenko, but he was willing to take unnecessary chances. Ensign Larin was a somewhat nervous, unstable man, always trying to prove that he was equal to and as skilled as the other two. He was good, except for the fact that he was jerky and inconsistent in his execution of acrobatic flying. Flying in 1917 was a risky occupation because aircraft engines were often unreliable and to take unnecessary chances could prove fatal.

On one occasion, Lieutenant Turenko was invited to demonstrate his flying skills at Gatchina. He took off in a Nieuport 17 and, after a brief warm-up, swooped the airplane into a wide climb upward, gradually banked to the left, gently rolled over, reversed his direction—up, over, and back to his starting point, all with the grace of a soaring bird. After executing most of the acrobatic stunts known at that time, he landed.

Within moments he was surrounded by an admiring group of pilots showering him with praise. "You're the greatest! That was the best Immelman we've ever seen!" someone shouted, "even better than the Frenchman's!"

"A couple of weeks ago a French flyer came to our school especially to teach us acrobatics," one of the students volunteered. "He performed perfectly—at the beginning. After his first act, he offered to demonstrate a landing directly from a spiral tail-spin. Unfortunately, he began to pull out of the spin a fraction of a second too late—his aircraft struck the ground and he was killed!"

Turenko listened to the story, apparently taking it as a challenge, and responded, "That's a stunt I'm sure I can do. Watch me take a turn at it!" He hopped into his trainer and was off. At an altitude of about 3,500 feet, he put his Nieuport into a tailspin, that is, he forced it to fall downward in a vertical spiral. When he approached the ground, he pushed the stick forward to force the nose downward to stop the rotation and make the airplane manageable. It stopped rotating, Turenko pulled the stick toward himself to land, but alas, he, too, was a fraction of a second late in doing so and had simply repeated his predecessor's mistake!

Two months after Turenko's death, Ensign Chepurin went to Gatchina. Of course, he was invited to demonstrate his acrobatic skills and, unbelievable as it seems, he repeated Turenko's performance—with the same fatal result. Why the administrators of the school permitted such risky stunt flying was beyond any line of reasoning. Two of our best pilots had been killed needlessly because no one in authority had made a move to prevent the tragedies.

Some pilots who had been to the front and suffered from war fatigue were sent to our school for additional training in Nieuports. Actually, they were nervous wrecks in need of rest and recuperation, not further training. At first, they were not eager to talk about their experiences, but later, when they had recovered sufficiently, some were willing to share their knowledge of aerial combat with us. However, we came to the conclusion that the information they offered us would be of little value in confronting the German "aces."

My flying proceeded normally, without incident, except for a few difficulties. One time while on a usual training flight, I was practicing a right vertical bank—a sharp right turn where the wings take a vertical position. It was a warm, sunny, windless day, the trainer obeyed my every whim, and I was happy as a lark when all of a sudden the airplane turned to the left and went into a barrel roll. It began to drop like a rock in the direction of a stone hedge surrounding our living quarters. Struck with the thought that the tail section must have broken off from the fuselage, mak-

The flight line at Odessa in 1917. These are Nieuport fighters.

Maintenance men remove a section of wing from the wreckage of Nieuport fighters at Odessa.

Alexander Riaboff flew this Nieuport 10 fighter at Odessa during the summer of 1917.

A group of cadets at the Odessa school stand next to a Nieuport fighter in one of the facility's tent hangars.

ing the airplane completely unmanageable, I knew that within a few seconds my machine would hit the hedge and that would be the end of me.

In a flash a short life of twenty-two years passed before me—my childhood, happy summer vacations, schools, friends, and my girl friend! And, as though to complete the picture, I envisioned my aircraft wrecked on the hedge. I saw my broken body and a crowd of pilots and crewmen running toward the scene of the accident. But in spite of this grim fantasy I felt no fear!

My dream ended when I noticed that my falling airplane was also rotating about the axis of the fall, and I said to myself, "Why, this is only a 'flat tailspin.' " I certainly knew how to handle that, so I pushed the stick forward to increase the angle of the fall to stop the rotation, then I leveled the airplane and landed. My expectation of death disappeared the moment I took control of the situation, although I must admit I felt somewhat uneasy for a time after setting foot upon the ground.

When I told my story to others on the field, one of the pilots said, "I saw you go into a spin, but I thought you did it intentionally and I attached no significance to it." In those few moments I had relived my entire life— had even died and in some mysterious way had been born again! Not a single student pilot was killed at the Odessa school during the time I was there. Nevertheless, there were several accidents. I witnessed two that could have been prevented had the administrators been on the job.

One student flew too close to the wires in an attempt to jump over them, hit them with his landing gear, wrecked the plane, and injured himself. Another student pilot took his friend, a civilian, for a flight in a two-seater trainer. He also attempted to jump over the telegraph wires at the end of the field with the same dire results. He wrecked his airplane, injured himself, and his friend lost a number of teeth.

One of our mechanics, a young, handsome, willowy Georgian from the Caucasian Mountain region, was well known for his ability to perform the famous Georgian toe dances. One day when I was seated in my trainer and ready to go, he came over to me and said, "Could you please make your flight as short as possible so as to give me more time to get to Odessa? I have an engagement to dance there." It was his duty to service my airplane before leaving the field in order to get it ready for the next flight. I was glad to oblige him, and returned to the field in half my usual time.

When I landed I joined a group of pilots. I then noticed that a student had gone to my airplane, but then I paid no further attention to what followed until I heard a sharp scream coming from that direction. I ran over to the airplane and saw the pilot on his knees in front of it attempting to do something.

"Help, help—please, someone help me!" he cried in desperation. The poor chap was stunned by what had happened. There, lying on the ground under the propeller, was our handsome Georgian mechanic. He had received a deep gash on the left side of his head, caused by the airplane's propeller, and it was obvious that he had been killed instantly.

The question was, how could this accident have happened? There was a hard and fast rule that a mechanic could not touch the propeller of an aircraft without permission of the pilot. We came to the conclusion that the mechanic, being in a hurry, had decided to turn over the motor in order to draw gasoline into the engine so it would be ready to start immediately. Somehow, the contact was closed, or the engine was hot enough to fire. Whatever the cause, we had lost a valued mechanic. To disregard a standing rule of procedure could prove fatal, as it did in the case of our handsome, talented, Georgian mechanic.

Life proceeded at a leisurely pace at our flight school, but the tempo of the Revolution began to increase by leaps and bounds. In Petrograd the Bolsheviks had seized power on November 7, 1917. Workers and sailors had taken over in the port city of Odessa and had begun to persecute all who opposed them. Right-wingers, including the well-to-do, former government officials, military officers, and the general class of intelligentsia, who made up the so-called antirevolutionary element, could no longer safely walk the streets in Odessa. Ominous rumors had reached us to the effect that officers had been arrested by bands of sailors from the cruiser *Potemkin*,* which was anchored in Odessa's harbor and was staffed by revolutionary forces.

The arrested officers were taken aboard, interrogated, and judged by a group of sailors. If they were found guilty of counterrevolutionary activities, the officers were thrown overboard into the Black Sea with weights attached to their legs and their hands bound. An alternate method of disposal was to toss them alive into the ship's furnace.

Other gruesome stories began to reach us. Once a diver, sent down to inspect the bottom of the *Potemkin*, encountered a sight that drove him mad temporarily. Shortly after his descent, he gave a signal to be pulled up. Upon removing his helmet he stood on the ship's deck repeating hysterically, "They are all alive, all alive, standing upright!" Other divers were sent down to investigate this grisly report and they found that the bodies surrounding the ship's bottom were afloat in vertical positions, being moved about slowly by the sea's current. To be sure, it was a macabre sight.

---

*During the 1905 Revolution, the crew of the *Potemkin*, a warship in the Black Sea fleet, mutinied. Later, the famous Soviet film director Sergei Eisenstein immortalized the revolutionary event in the film *Potemkin*.

Alexander Riaboff (center) stands with a group of pilots and mechanics at the Odessa school in the summer of 1917.

Officers, cadets, and mechanics gather around a Nieuport fighter. On this day the school received a group of officers who had come to inspect the facility.

The wind-swept, undulating field at Odessa caused frequent landing mishaps. Alexander Riaboff (second from left) stands next to a Morane fighter that had nosed into the field in a morning exercise.

Flight instructor Captain Alexander Larin (center) poses with cadets at the Odessa school. Captain Larin gave instruction on Nieuport 10s. Cadets at Odessa first qualified in Moranes and then were given instruction on the more advanced Nieuport models.

As conditions in Russia grew steadily worse, those who opposed the Revolution became terror stricken. Their lives were in constant danger, and they had no apparent means of escape. Some actually waited anxiously for the arrival of the German Army, which was moving slowly, without opposition, into southern Russia. The Russian Army had disintegrated and the Bolsheviks had declared a policy of fraternization with the Germans, but some people did not trust the German Army's benevolence and moved to northern and eastern sections of the country.

A breakdown of discipline finally reached our school, leading to its complete disintegration. People had begun to leave under various pretenses. One of the first to depart was our doctor, who had received a telegram from his wife, living in Siberia, saying that she was seriously ill and wanted him to come to her immediately. Although the doctor's excuse for leaving was suspect, our commander granted him a furlough for one month, knowing full well he would not return. Some officers adopted the same procedure to gain furloughs and the school's personnel began to shrink visibly.

One day, to our great surprise, we were notified that on the coming Sunday, there would be a meeting of the flight school's personnel, which would also be attended by a group of "comrade sailors" from the *Potemkin,* for the purpose of discussing certain misunderstandings and difficulties occurring, supposedly, between the commanding staff and the rank and file. The comrade sailors were to be the arbitrators to help settle the troubles allegedly brewing. This startling news promised to be an event pregnant with horrifying possibilities!

Valerian Przhegodsky and I discussed the wisdom of attending the proposed meeting and decided to risk it, but, we agreed, that under no circumstances would we permit anyone to force us to board the *Potemkin.* We went to the meeting that Sunday, each armed with a concealed weapon. Almost the entire personnel in our school, a large representation of comrade sailors, and *Potemkin* officers were on hand. There had been no prior registration of participants nor had documents been presented authorizing the comrade sailors to arbitrate the so-called "difficulties." Ironically, anyone was welcome to join the meeting, to vote, and to express himself fully in complete freedom.

After a few short preliminary speeches, the meeting was declared open for discussion of the trumped-up charges. I was the first one to ask permission to speak. "Comrades," I said, using the new classless terminology of addressing one's fellows, "I have been a student at this school for several months and, so far as I know, peace and harmony have always existed between the commanding staff, officers, and the rank and file. Had there been trouble, don't you think we could have adjusted and compromised our differences amongst ourselves? It seems to me that you, as strangers, unfamiliar with our school lifestyles and traditions, should not become involved in our affairs. You are in no position to interfere or act as judges in disputes amongst us."

"We have the right to judge you," shouted a sailor some distance away from me. "You are counterrevolutionaries—all of you officers!" One by one the sailors rose to their feet and pointed accusingly at the officers, yet they were unable to name a single name. At any moment the melee could have resulted in a shooting foray. Worse yet, we could have all been taken forcibly to the *Potemkin*!

An elderly man, a longtime revolutionary and a former officer in the tsar's Imperial Navy, arose to speak. His timing was perfect. "My dear Comrades, permit me to take a few moments to remind you of some facts from Russia's history. First, I must remind you of the role played by officers and intelligentsia in the long struggle for freedom in our country. Do you know to what class the majority of army and navy officers belong?"

"To the bourgeoisie!" someone shouted.

"No comrade, you're wrong! You assume they belong to the upper strata of society, but that is not true. Most of them come from the lower strata, and only the top brass, the generals and admirals, are members of high society and the nobility. You'll find that revolutionary movements, including the Decembrist Insurrection in 1825 against Emperor Nicholas I, were supported by the officer class and intelligentsia, along with the common people. A total of 492 revolutionaries [a gross exaggeration] were hung and many hundreds were sent to Siberia for life when the December, 1825, insurrection failed. Remember this, every revolutionary movement was fomented by the intelligentsia and officers, who were all severely punished or killed when apprehended by the tsar's forces.

"It's time we turned our attention to the present Revolution and acknowledge that we have thousands and thousands of officers and intelligentsia working and fighting with us in a common cause. Our goal is to establish a new order in Russia with freedom and equality for all! Returning to the present situation, may I stress the fact that no individual officer at the school has been accused of any counterrevolutionary activity. It's time to take a vote of confidence and cease making unsubstantiated accusations. From now on, let's work together for the success of the Revolution and for a bright, happy future!"

Resounding applause followed his dynamic speech, and as soon as it died down, our school's sergeant, at the secret urging of our commander, announced that the rank and file had no complaints against the officers in Odessa's flight school. His statement paved the way for the unanimous

An ambulance was available in case of any serious accident at Odessa. The pilot who crashed in this trainer was only slightly injured.

Riaboff's Polish friend, Valerian Przhegodsky, overturned this Nieuport 10 on the "washboard" airstrip at Odessa. Riaboff (right) stands next to the overturned aircraft with Przhegodsky (left) and an unidentified mechanic.

Cadets, including Riaboff at left, point in jest at a cadet named Ivanov, who had just crash-landed his Morane aircraft.

Another shot of the Nieuport 10 crashed by Przhegodsky, who is on the ground, leaning on the wing. Riaboff is seated on the ground at right with his hands clasped on his knee.

vote of confidence that followed, and vindicated the school's officers of any implication in counterrevolutionary activities. Thereafter, the meeting was adjourned.

I felt a tap on my shoulder and turned about to find my commander, Captain Chekhutov, with a look of urgency on his face. "Get the hell out of here as quickly as possible without drawing any attention to yourself," he whispered hoarsely giving me a shove. I wasted no time obeying his order. In retrospect, I have often contemplated what might have occurred at that meeting had I not disrupted it with my speech, which diverted the sailors' attention from their original plan to accuse, arrest, and try their victim officers, culminating in a fatal trip to the *Potemkin*.

After passing our final flying examinations, some of the students left Odessa for parts unknown. The armed services, including our school, had deteriorated to the extent that there were very few specific orders given to anyone as the Germans advanced slowly along the entire front.

Captain Chekhutov called me to his office. "Ensign Riaboff, I have an order to send you to the front."

"Yes sir," I replied, snapping to attention. "When do I leave?"

"You're not going anywhere!" he responded emphatically. "You remember Lieutenant Smirnov, who was ordered to the front a month ago? Well, he was halted by a crowd of soldiers near the end of his journey and was asked, 'where are you going and why?' He told them that he was on his way to the front to join an aviation division as ordered. With that, they killed him because, they said, he intended to fight their German 'comrades' and continue the war. We do not want you to suffer the same fate. I want you to join a special detachment organized at the school to fly reconnaissance over the vicinity of Odessa." So two or three times a week, I flew over the city at an altitude of about 5,000 feet, but I never encountered a German airplane.

News from the front continued to be bad, and, no wonder, our army had lost its fighting ability completely. The German Army was moving slowly to the west and south unopposed, and it was obvious that Odessa would be occupied by the Germans very soon.

My friend Przhegodsky and I discussed the occupation and what might happen to us in that event. His opinion was that we should wait until the Germans arrived, then we could go to Poland and join the Polish Air Force. I was against his plan, as I did not trust the Germans to treat us well. It would be more logical to assume that they would detain us as war prisoners and send us to Germany. It was too much to expect them to treat us with consideration. Furthermore, Przhegodsky's idea of joining the Polish Air Force might be a reasonable possibility for him, a Pole, but for me, a Russian, it was inconceivable that I would be accepted, because the

Poles consider Russians to be inferior and, I may add, we Russians have the same opinion of them.

Although Imperial Russia no longer existed, the Communist government was not yet fully formed. No one knew his duties or his rights. *Ad hoc* groups took it upon themselves to appoint commissions to assume governmental functions, and no one thought to question a commission's authority or rights. So as to avoid being taken as a prisoner of war, the most sensible plan for me was to return to Moscow as soon as possible. To hesitate any further was not feasible because of the chaotic times. I knew that the trip to Moscow would be fraught with danger, so to travel in my officer's uniform was out of the question. I decided, therefore, to masquerade as an aviation mechanic—and I would carry documents to prove it.

At that time a student's commission in Odessa acted on behalf of all cadets and it assumed authority to demobilize them. I went to the commission's office and presented papers showing that I had attended the Institute of Communications in Moscow. The commissioner examined them and gave me a document stating that I had been demobilized and had the right to travel anywhere I chose. I showed Captain Chekhutov my right-to-travel document and told him I intended to return to Moscow. He agreed with me that it would be the wisest move I could make in my present situation.

Przhegodsky and I helped ourselves to the school's typewriter and the two of us worked laboriously to prepare my fake documents showing that "Alexander Riaboff, aviation mechanic" had been demobilized and was returning to Moscow. The sergeant in charge of the school's office had noticed the absence of his typewriter and went in search of it. Just as we had finished with it, the sergeant, a timid chap, came to our room and inquired hesitantly, "By any chance have you seen, or a-a-a I should say, did you see anyone take the typewriter from the office?"

"Oh yes," I replied, "We're through with it. Here it is—just help yourself." He looked at us questioningly, but left without satisfying his curiosity.

Landing accidents were always an occasion for posing for a photograph. Lieutenant Kudrin, who broke the landing gear on his trainer in this mishap, is pictured with a number of the men from the ground crew at Odessa.

An Austrian prisoner of war is shown at right in this photograph. The Odessa school, as well as other airfields, employed volunteer POWs as a means of meeting a shortage of skilled aviation mechanics.

5. Under the Red Banner

Alexander Riaboff returned to Moscow in February 1918. His native city had become the besieged capital of the Bolshevik regime. Riaboff's family had survived the revolutionary events of the previous year, but they faced the threat of famine. Life in the city was bleak, punctuated by violence and terror, and ruled by commissars intent on exacting "revolutionary justice" and mobilizing the resources of the beleaguered city to support the Bolshevik regime. There were dangers for those who were associated with the old regime either by social class or, in Riaboff's case, military rank.

In May 1918, civil war broke out as various groups arose to challenge Bolshevik authority. Riaboff participated in one of the first major battles of the civil war, the struggle for the control of Kazan on the Volga River. Here the young pilot found himself in the Red Air Fleet, at the time assisting Leon Trotsky in the Bolshevik drive to capture the ancient city of Kazan from the White forces. It would be during this conflict that Riaboff determined to defect to the Whites. His account of these days is of great historical interest because it sheds light on the embryonic Soviet Air Force and its first commander, Konstantin Akashev.

The Bolsheviks successfully defended Moscow in 1918, and captured Kazan from the Whites in September of that year. The civil war, however, would continue for two more years as the Whites endeavored to overthrow the Bolshevik regime.

■　　■　　■　　■

L ATE IN FEBRUARY 1918, I SAID GOOD-BYE to Przhegodsky, Captain Chekhutov, and the remaining personnel. When I departed, spring had arrived. Lustdorf was already warm and sunny and some people had begun to arrive from Odessa to enjoy the beaches and the general serenity of the village. What a contrast to Moscow in the middle of winter!

A DH 4 bomber mobilized for service in the Red Air Fleet. This particular aircraft was used in the Kuban area, in the Caucasus region, against the Whites led by General Denikin.

In preparation for my trip, I procured a long soldier's overcoat, stopped shaving for a few days, then packed a few essential items in a lightweight box salvaged from a wrecked airplane, and I was off. By this time the school had stopped functioning altogether.

Upon reaching Odessa, I had planned to board a train for the city of Nikolayev, situated on the estuary of the Bug River, but I was advised that the Germans were threatening to cut off the railroad and I could be more certain of getting to Nikolayev by boat. So I boarded a small, ancient, over-loaded steamship crowded with passengers going to Nikolayev. It was nothing short of a miracle that we reached that city without incident.

Although Nikolayev lies only 100 miles east of Odessa, the climate is markedly different. Snow still covered the streets of the city and it was very cold. On the day of my arrival I found a train heading for Moscow, loaded with soldiers returning to their homes. I joined them in a boxcar, and the train began to move slowly northward, but it was forced to change its course to an easterly direction when the engineer received a message that the Germans had threatened to cut the railroad line to Moscow. Two more times the train was forced to change course as we moved slowly, spending long intervals at stations along the way. No one checked the passengers, or asked for documents, or even tickets.

We were enjoying a certain feeling of freedom when the train stopped at a large station and some men entered the boxcar. One of them, presumably the leader, asked, "Do you fellows have any officers riding with you?" Because I was standing in front of our group of passengers I became the center of attention for the intruders. I was unwashed, unshaven, and unkempt in appearance, just one figure in a motley band of rank-and-file soldiers. I behaved myself in a cool, casual manner, but it seemed to me our inquisitor was taking a particular interest in me and was about to question me when a voice from an upper bunk said sarcastically, "Well, they wanted to promote me to officer's rank, but my mug wasn't right for the job!"

This remark infuriated the man, who immediately leaped to the bunk, took one look at the fellow, and remarked with disgust, "Oh, to hell with you!" The group in search of officers then walked out of the boxcar convinced they would not find an officer in this paltry crowd. Once again, I breathed a great sigh of relief. Later, I was told that our visitors were in search of officers traveling east to join the White armies operating north of the Caucasian Mountains. A few days before, several former army officers had been caught and had been summarily shot.

Finally, our train reached the large industrial city of Kharkov, where all the passengers were ordered to get off. It was evening and I was exhausted, dirty, hungry, and confused, with no idea how to improve my

A close-up of Alexander Riaboff (center) at Odessa in the summer of 1917.

condition. Suddenly, I recalled that there was a branch store of my father's firm in Kharkov, and its manager, a man named Pankratov, was my father's close friend. I found his name in the directory and telephoned him. He remembered me. I told him that I was en route to Moscow, and then boldly asked him if I could spend the night in his home. "Please, come immediately," he replied warmly.

My host greeted me when I arrived at his apartment an hour later. I detected a look of apprehension and disappointment on his face at my appearance, but he let me in. After the usual exchange of greetings, I asked my host and hostess for permission to wash, shave, and improve my appearance as best I could. When I reappeared, their apprehension was gone and we sat down to a wonderful repast, after which we talked late into the night about the Revolution, the future of the war, and the fate of our beloved country. We agreed that Russia faced extremely difficult times ahead. There seemed to be no way to escape the anarchy, violence, and hunger that would befall us as the disintegration process continued. The paramount question was whether Russia could continue to exist at all.

The next day, the Pankratovs told me that no one was permitted to go to Moscow without a permit issued by a special commission. When I reached the office that issued travel permits, I was confident. Having been a resident of Moscow since birth, I assumed that I would not have any difficulty obtaining permission to rejoin my family in that city, now the new Bolshevik capital. I explained my situation to the commissioner and received a firm "no" from him. "There is no place to live in Moscow," was his final remark.

I argued with him at length, but to no avail. In fact, I returned to him every day for one week until Sunday, when he was replaced by another man. The new man listened to my story and granted a right-to-travel permit to Moscow without any opposition whatsoever. During my week of waiting, the Pankratovs treated me with the love and kindness reserved for their own son. I was deeply touched when Mr. Pankratov invited me to remain with them and to take a position in the firm.

I thanked them sincerely for their kind hospitality and said, "It is my duty to return to my family in this time of trial, and I am sorry to have to leave you most wonderful people." When I arrived at the station, I learned that the train would be delayed for several hours. The station was crowded with people, mostly soldiers heading for home. It soon became obvious to me that my permit, which had been so difficult to procure, was useless.

I realized that my main problem would be boarding the train! No one was even selling tickets! It would be an "every man for himself" situation. Finally, when the train arrived, I was shocked at the unbelievable number of people packed into it, on top of it, under it, and even between the cars and on the steps. The waiting crowd stormed toward the train, only to be met by the mass of humanity rushing from it. The result was pandemonium. A current of bodies pushed me onto the steps. I grabbed a handrail with my left hand, put one foot on the bottom step, and clung to my suitcase with my right hand. I remained in that position for several minutes.

Realizing that I could not hold on much longer, I decided to swing my suitcase up onto the heads of the people standing on the steps in front of me.

"What the devil are you doing?" shouted an exasperated passenger, lacing his remark with many unprintable expletives.

"Comrades, please pass my suitcase forward," I said pleadingly. "I can't stand on the last step with only the one foot, hold onto the handrail with one hand and the suitcase with the other another minute."

My suitcase disappeared as the crowd steadily pushed forward, moving very slowly into the car. I gradually advanced from the bottom step to the top, and finally, to my relief, reached the car's entrance where the wind, snow, rain, and freezing cold would not paralyze me. We were packed in like sardines in a can but, for the time being, I was at least safe from the elements. Then the long-awaited bell and whistle signals sounded and the train jerked forward, only to be immediately brought to a halt.

There was a great deal of commotion and hollering outside and someone yelled, "Some poor devil has fallen under the wheels!" It was a soldier, we were told, who had been standing between the cars on the connecting links. Another attempt to move the train resulted in a second death for the same reason, but after that mishap we were on our way without further incident. I finally reached the inside of the car, retrieved my suitcase, and sat on it.

By midnight I was dead tired after spending fourteen hours waiting for and finally boarding the train. Finding a place to sleep seemed an impossibility, but luck was still with me. I spotted a narrow, empty baggage shelf made of heavy netting high above a window. Plotting a way to reach it, I began to move.

"Hey, you, where do you think you're going? There's not enough room to stand and you want to go somewhere!" a soldier shouted.

"If you'll just wait a minute, I won't bother you for long and you'll have a little more standing room too," I said as I pushed forward.

Within a few moments I was comfortably lodged in the netting and fast asleep. The next day I awoke to find the train less crowded, although we would not arrive in Moscow until evening. Moscow was bitterly cold, dark, and blanketed with snow. I had been away less than a year, but I sensed that the world had changed completely and I felt a sudden dread of what I might encounter. I hailed a droshky, or horse-drawn cab, waiting at the station for hire. I climbed in, gave the driver my address, and in a short time we were within a stone's throw of my home. Suddenly, I noticed a large and strange-looking object lying right in the middle of the street. I was shocked to discover that it was the carcass of a horse. Stray dogs had ripped open its belly and its insides were gone, proving that it had been lying there for some time.

The Moscow I had left earlier would have disposed of the dead horse before such a dreadful thing could have occurred. "What about this dead horse left in the street?" I asked the driver. "Oh, that's not at all unusual for these times," he replied with a shrug of his shoulders.

We stopped in front of my home and I asked the driver how much I owed him. "I don't want your money," he snapped.

"Well then, what do you want?" I asked.

"A part of what you have there," he said, pointing to my suitcase. "I know that you are returning from a trip to some village where you bought potatoes, butter, flour, and other stuff I can use—that's what you have in that suitcase. Money isn't worth a damn now, but food is scarce and worth plenty!"

Moscow under the Bolsheviks in 1918, as marchers pass through Red Square. St. Basil's Cathedral (left) and the Spasskiy gate to the Kremlin (right) are clearly visible, along with a balloon used as part of an air defense system set up by Bolshevik authorities.

"I'm sorry, fellow," I replied. "There's no food in that suitcase because I've just returned from the front." His disappointment was apparent as I slipped him some money to appease my conscience. "Thousands of people leave the city every day to get food in the provinces. They exchange household articles, clothing, and jewelry for food." His parting words were "You can't buy food in Moscow!"

To think that Russia at one time had fed Europe and now she was headed for a famine! Yes, Moscow certainly had changed!

As I climbed the stairs to our third-floor apartment, I noticed that they were dirty and there were piles of firewood stacked on the landings and clothes had been hung to dry on lines strung-up overhead—an unusual sight, to say the least. My heart beat rapidly with excitement as I pressed the doorbell and waited. The door opened and I saw my eight-year-old sister, Tanya, who stared at me unbelievingly for a moment, then screamed at the top of her voice, my nickname, "Sanya!" For some unknown reason, she promptly ran from the room, seemingly frightened, as though she had seen a ghost.

This visit was an emotional reunion with my parents and my sisters.

The unexpected meeting sparked a great outburst of joy for all. War and revolution had brought separation and, for my family, real privation. Combined with these circumstances was a sense of uncertainty about the future. For a moment these anxieties were overcome by the joy of our reunion. We were thankful just to be together—the prodigal son had returned!

It took several days for me to get accustomed to being with my family, in my old room and with all the comforts of home. Everyone, it seemed, had news of various kinds to share. In long conversations, news was exchanged on the impact of the Revolution on our lives and the lives of our friends. Father told me about the drastic changes that had taken place in his firm.

"The employees," he said, "have taken over and the owners have lost all their rights to the business." He described how a meeting had been called by the revolutionaries to reorganize production and to elect new management. One of the first things the workers demanded was a raise in salary. They also wanted the work week shortened drastically. My father told them that there was no reserve of capital and it would be impossible to meet their demands. One worker suggested that money could be raised by selling raw materials on hand. My father replied "how are you going to run a factory without raw materials?" After that exchange, the meeting was adjourned without any firm decisions having been made.

Soon after the meeting my father was informed that the workers intended to elect him manager, but he was frightened at the prospect of running the business in accordance with the workers' demands. "I decided to go to the summer home and wait for someone else to be elected," he said, with a tone of regret in his voice. Thus ended his forty-five years of service with the same woolen manufacturing firm. Eventually, he found a position with the government agricultural cooperative as an accountant.

During my short stay in Moscow, I managed to locate two of my old school friends, Sergei Popov and Michael Zenichev. Popov had stubbornly continued his studies at the Institute of Communications and had managed to pass all the examinations required of him to avoid serving in the army. Zenichev had entered the Institute of Technology in Moscow, but later left school to volunteer in an artillery officers' school, where he received his commission and served until the army disintegrated. Time had altered our circumstances and outlook. I somehow felt that our close friendship had deteriorated because our interests and occupations were entirely different.

My old friends urged me to return to the Institute of Communications and resume my studies. My reply was equivocal and half-hearted. I did visit my alma mater and found it almost devoid of students. Everything appeared to have changed—and for the worse. I located my old desk, now occupied by a thin youngster wearing thick glasses, working at his drafting. Compared to him, it seemed, I was sophisticated, a man of the world, so to speak, and I concluded that it would be impossible for me to resume the role of a student and work next to this youngster. "No," I said to myself, "I am a pilot first and foremost. I cannot leave aviation and return to my old school."

There were other acquaintances to visit. I decided to call on Colonel de Savari and his family. He had been a colonel in the Russian Army and a professor at the Moscow Military School. Now this former officer in the tsarist army was teaching officers destined for service in the Red Army. Life obviously had taken a turn he could not have anticipated in former times. Earlier, I had inquired about the origin of the name de Savari, as I had always wondered about the family's French ancestry. One of the colonel's daughters produced a book describing the family history. The de Savaris were descended from a French duke. When I asked for more particulars, I was told that during the Napoleonic invasion of Russia in 1812, one of the de Savaris had been captured and, after the defeat of the French, had chosen to remain in Russia. His descendants traditionally served in the Russian Army.

For this reason, I considered it a privilege to visit the de Savari family; I had always enjoyed my conversations with such intelligent and cultured people. Mrs. de Savari, her two daughters, who were about my age, and her somewhat younger son, met me at the door that day. They greeted me warmly and told me that the colonel had not arrived home as yet, but they assured me I was welcome to come in and wait for him.

He came home unusually late that evening and after exchanging greetings, he explained the reason for his tardiness: "My classes were finished for the day. I boarded the streetcar for home, as usual, and while riding along I observed a worker sitting across from me who was holding a large sack of potatoes on his lap. I also observed a hole in the sack of sufficient size for an average potato to pass through. I quickly deduced that when the man disembarked from the streetcar he would have to carry the sack over his shoulder, as it was quite heavy. Furthermore, there was a good chance that some potatoes would fall through the hole onto the soft snow unnoticed by the man.

"I decided then and there to prove my logical deductions, and when he got off the streetcar I followed him. For some distance nothing happened, and then, to my extreme delight, a potato fell out. I picked it up and continued in his footsteps until—Lord be praised—another potato fell onto the snow unnoticed by him! I followed him further for a considerable distance, but in vain."

With a broad grin on his face and genuine pride in his voice, he said, "Now, my dear, to prove to you and others present that my story is not

This Il'ya Muromets was photographed at the Moscow Military Airfield in 1916. During the Bolshevik Revolution, these aircraft were mobilized for service in the new Red Air Fleet. Note another Il'ya Muromets in the hangar in the background.

fabricated, I present you with two potatoes!"

One would have thought that he had presented his wife with two Fabergé Easter eggs. This entire episode shocked and disturbed me deeply. To think that a professor, an army colonel and a cultured man, had followed a worker carrying a sack of potatoes for at least a mile hoping to acquire a potato or two. This anecdote indicated the desperate circumstances of Moscow under Bolshevik rule. The food situation in the capital of the new Soviet Russia was bad, but I thought how much worse conditions must be in other cities. Imminent starvation seemed unavoidable without help forthcoming.

Our family was fortunate in having relatives who lived in a nearby village and raised enough produce to share with us. During my stay in Moscow I met many more old friends and acquaintances, all of whom seemed lost and insecure. Their thoughts were centered solely on the acquisition of food, on the unpredictable future, and on elemental tasks of survival in revolutionary Russia. I, too, had begun to share their feelings of loss, and did not know which way to turn.

Shortly after arriving in Moscow, I saw a Soviet Government order: "All pilots must appear on Monday, April 10, 1918, at the Moscow Airport for registration." For the Bolsheviks, human life was expendable during these terrifying times and the failure to comply with such a government order could lead to very unpleasant consequences. For this reason, I

The flight line at Moscow in 1918 shows the motley collection of fighters mobilized by the Bolsheviks to defend the Revolution. Composed mostly of Nieuports, these aircraft were used against the Whites at Kazan in the summer of 1918. Some of the fighters still display the old Imperial roundels.

decided to register. I have to admit that I was also curious to discover how many and who of the pilots would show up.

Among those who complied with the order was Captain Chekhutov, my former commander of the Odessa flight school. He was joined by many other pilots who came to register in response to this decree. Despite the fact that the gathering was fairly large, it was a silent and subdued gathering. Under normal circumstances there would have been a noisy, happy meeting, but under Bolshevik rule everyone was reserved and emotionless. As a matter of fact, we were apprehensive of one another; there was no way to be certain how anyone really felt about the Soviet Government. There was always the fear, especially for former officers, of being denounced as a counterrevolutionary. A gesture or a word might indicate one's political bias, which could be fatal.

Following the registration, there was a meeting. This session was led by a prominent member of the Communist party, not a pilot by profession, but a man appointed to supervise aviation activities for the Bolsheviks. No one doubted his power to shape our lives. Yet, his speech was intelligent, his manner mild, and he did not threaten us in any way. He simply invited us to cooperate with the Soviet Government, promising us full cooperation on its part. No visible reaction to his promises showed in the faces of those who had listened to the speech. There was no way to judge the sort of impression he had made on the group. The assembled pilots, it seemed, had decided to wait and see how they would be treated.

My own impression was that we would be better treated by the Bolsheviks than most military personnel because we possessed special skills. Simply put, the Soviet Government needed pilots. Ensign Golov, a former Gatchina student, and I left the meeting together, walking toward the center of the city. Along the way we discussed freely our separate experiences since we last met. Although each of us was politically neutral for the most part, we expressed our opinions openly now that we were alone. Fearful of what fate might be awaiting us, we agreed it would be best for us to "wait and see" also.

As we were passing a hotel, Golov said, "Let's call on some pilots I met a few days ago who are staying here." I thought it was a good idea; the visit might provide an interesting interlude. So we climbed to the third floor, found their room, and knocked on the door. There was no answer, and when Golov knocked again we heard some sort of commotion inside. A voice called out, "Who's there?" Golov replied, "It's Golov and an old friend of mine, a pilot." The door then was opened, somewhat cautiously I thought, and three casually dressed men stood before us. Golov introduced them to me. Two of them had names unfamiliar to me, but one was well known as belonging to a pilot who had brought down several German planes.

After a half hour of polite, but guarded conversation, I had the feeling we had interrupted whatever activity they had been engaged in and that we were not quite welcome, so we left, and continued to walk to the city's center.

"Somehow, I got the impression those fellows might be conspirators and we caught them in the act of preparing to do something secretive!" I prophetically observed.

Sometime later I was not surprised to learn that the three pilots had gone north to Archangel, a city on the White Sea, which had been seized by former Imperial Army men with the aid of a small British detachment. True enough, at the time we encountered them they were making plans to join the Whites, a dangerous pursuit, which could have resulted in their instant deaths by a firing squad if they had been discovered by a Bolshevik sympathizer.

Later, Chekhutov and I met on several occasions to discuss our futures. At one of our meetings, he proposed the idea of organizing a new advanced flight school, on a smaller scale than the Odessa flight school, of course, but one that would be located in a small town just outside of

Moscow. We agreed that this project would allow us some time to organize our thoughts during a period when our country was gripped with a vast political upheaval.

Chekhutov took it upon himself to conduct the negotiations with the Soviet aviation staff. I offered to assist him in whatever capacity he requested. Within a very short time, the aviation staff gave us permission to proceed with our plans, which would require the mobilization of aircraft, equipment, fuel, a technical staff, not to mention food and other incidental necessities—all of which were scarce. Furthermore, no one in the present Soviet Government necessarily had any knowledge of our real requirements, let alone where they might be found. It was my assignment to take charge of the supplies problem, which I accepted with all my youthful enthusiasm.

Eventually, I was able to locate several Nieuports. I bargained strenuously to acquire these aircraft and finally received a promise that five or six of them would be delivered to us. They arrived in due time and were placed in boxcars set aside in one of the railroad yards. I also managed to obtain gasoline, lubricating oils, tools, and many other essential items. All were very difficult to locate, particularly gasoline. Finally, we received most of our necessities in reasonable quantities, including two machine guns with a supply of ammunition for training purposes.

In the meantime, the Soviet authorities gave Chekhutov the go-ahead to establish the flight school near the city of Samara (Kuibyshev), on the Volga River, where the railroad from Moscow connects with the Great Trans-Siberian Railway. An aviation school for beginners, under Captain Ivkov's command, had been operating there, so the staff decided that both schools could work together with either Ivkov, Chekhutov, or both men, in charge. This particular issue could be decided later. It was understood that Chekhutov would lead the advanced flight-training program.

At last everything we needed was loaded and ready to go to Samara, but before we could leave, it was necessary to obtain a permit from the Moscow Soviet Commissariat of Transportation. I went to the commissariat's headquarters, located in a former privately owned palace, and spoke to the administrator. He proved to be one of the Communists who had come to Russia from abroad after the 1917 Revolution and was barely able to speak Russian. When I explained our need to obtain a permit, he promised to issue one to us without delay.

Political instability at this time had intensified because of the growing opposition to the Reds. These groups of anti-Bolsheviks called themselves "White Armies" and were putting military pressure on the "Red Armies" in the south, the north, and from the Ural Mountain region in the east. I said to Chekhutov, "I recommend that we leave Moscow as soon as

A Nieuport fighter with Red Air Fleet markings.

possible because of Russia's increasing instability." To my surprise, Chekhutov, usually a cautious man, responded, "Oh, what's the hurry? Life is pleasant here in Moscow. Samara can wait for us a little longer."

In due time, I learned that Chekhutov had fallen in love with an attractive woman in Moscow and was in no hurry to break off the relationship. After some time had passed, Chekhutov told me that he was ready to go to Samara, and that I should obtain the permit for our departure the following Monday. We would leave Moscow without delay. On the next Monday, I arose early, sat down to breakfast, opened the morning newspaper, and, to my great dismay, read the following: "THE WHITE ARMY IN SIBERIA OCCUPIES SAMARA." The newspaper went on to report that "Soviet troops have been sent to clear Samara of the 'White bandits' and we hope this operation will take only a few days."

Our plan to establish the school in Samara had received a definite setback. Nevertheless, I decided to go to the commissar of transportation to ask for the permit. He confronted me with the morning headlines, but assured me that as soon as the situation was rectified, the permit would be granted. As the weeks dragged on, the White Army still held the town, so there was nothing to do but select another location for the school. After studying a map of the Moscow region, we selected a small town called Alatyr, which lies about 350 miles east of Moscow and 70 miles southwest

These Red Air Fleet fighters served at Kazan in 1918–1919.

of Kazan. I was sent to inspect the town and its surroundings to determine the suitability of the location for our needs.

When I boarded the train for Alatyr, I found it packed to overflowing with men going to the villages to buy food. The train proceeded slowly, stopping at each station, sometimes for long periods of time. At one of the stops, a great number of people got off, and when I looked out of the window I saw a man walking by slowly while counting the money he held in his hands. Suddenly, another man, going in the opposite direction, grabbed the money and started to run away. I wondered how the thief, engulfed by people, could expect to escape with the money. "Only a fool would attempt such an act," I mumbled to myself.

"Hold him, hold him, he stole my money!" screamed the astonished victim. The robber was caught immediately. He then was brought back to his victim and compelled to return the money.

One would have thought the matter closed at that point, but such was not the case. The men who had caught the thief began to beat him mercilessly, which proved to be an open invitation for the crowd to join in. I was shocked and sickened by the cruelty that unfolded in front of me, and I realized that the robber would be killed if something were not done immediately.

Impulsively, I ran out of the car and screamed at the crowd at the top of my voice, "Stop! Stop beating the man!" For some unexplainable reason they obeyed me and quieted down, enough so that I could speak my mind: "It's foolish and cruel to kill a man over a few rubles. . . . I'm willing to pay the victim of the crime an additional sum if you'll let the thief go."

No sooner had I made these remarks than a man in the crowd pointed his finger accusingly at me: "Why would a stranger want to protect this thief?" he sneered. "Comrades, how do we know whether or not he and the thief are in cahoots?" I glanced hurriedly about, and when I saw the unfriendly, menacing eyes staring at me, I sensed that the crowd was out for my blood, too. Fortunately, at that moment, a calm, assured voice coming from somewhere close by announced, "Oh no, comrades, look at this fellow, then take a good look at the thief—they are birds of a different feather. The thief operated alone!"

I could hear murmuring among the men in the crowd, evidently discussing some new strategy or saying they agreed with my benefactor. At any rate, they stopped beating the thief and took him to the station. My savior, a middle-aged man, came over to me and said excitedly, "Leave quickly, young man, before the crowd gets its chance to reconsider. They're unpredictable."

"Thank you, sir, I'll do what you say. . . . I'm truly grateful to you. . . ." I then moved hurriedly to another car.

I waited and watched to see what would happen to the thief. The crowd began to increase, surrounding the station, pushing and shouting persistently. "Give us the thief . . . we'll show him what's what . . . surrender him to us!" Just then, the station bell rang, indicating that the train was ready to leave, followed by the train whistle, which signaled that it was about to move. The crowd of men ran, boarded hurriedly, but the train did not move. They understood immediately that this had been a ruse to distract them. Some laughed about it, but not everyone found it amusing.

Gradually, the crowd moved to the station again, surrounded it, and became more and more insistent, shouting: "Surrender the thief! Surrender him to us now!" As their demands became increasingly threatening, another bell rang and the engine whistle suddenly pierced the air, again interrupting the mob's potential violence. Everyone ran and boarded the train again, but it did not move. Laughter rippled through the crowd as before and, what is more, this game was repeated two or three times until the people simply laughed at the repeating bell and whistle signals. As the mob increased and their demands turned into threats, I heard someone shout, "Come on fellows, let's tear down the station!"

I sat in the train watching this mad scene, hardly able to contain myself. I felt completely frustrated and useless, as I dared not interfere to help bring these wild men to their senses. All of a sudden, I saw the thief emerge from behind the station, running from his pursuers as a rabbit from hounds. He dashed across the clearing toward the forest only seconds away from the blood-seeking mob. Swallowed up by the dense forest, both the thief and his pursuers disappeared from sight. After several minutes passed, which seemed like an eternity, I saw the pursuers returning to the train, obviously the victors.

As they passed by my window, I heard one man boastfully describe his conquest. "He was lying on the ground when I caught up with him. I had a large rock in my hand and I let him have it hard on the head. That was the end of him!"

"Oh God," I thought, "how cruel and merciless a mob can be. Their thirst for blood makes them oblivious to the sanctity of life!" Did this incident portend a bad omen for the so-called "bloodless revolution?"

In the summer of 1918, I found Alatyr to be a small, provincial, sleepy town with a few thousand inhabitants. Many Russian villages resembled Alatyr at that time. On the town's outskirts, I discovered some large, well-constructed barracks built by Austrian war prisoners, and many of them stood empty now. In front of the barracks was a level meadow, too small for an airfield but large enough for small aircraft. Surrounding it was a large drainage ditch with a line of posts supporting electric wires hanging along them. It would be easy to fill in the ditch and move the posts and wires, I decided, thereby converting the meadow into a safe and convenient landing strip. I also concluded that in the future the airstrip could be increased in size to become a real airport.

When I returned to Moscow, I reported to Chekhutov, and soon the authorities decided to establish the school in Alatyr. Our permit was procured, the train accommodations made, and we were ready to depart. My family seemed pleased when I told them that I was stationed just a few hundred miles from Moscow and I would return for a visit within a few weeks. We parted casually: "I promise to return before you have time to miss me," were my parting words to mother, and I believed that to have been the truth. How could I have known then that I would not return to Moscow until 1960, as an American citizen, after an absence of forty-two years!

Once we reached Alatyr we plunged into our work. In the beginning, the school's personnel consisted of Chekhutov, the head of the new school, his brother, who was in charge of equipment and acted as secretary-treasurer, and Golov and I, who served as pilot-instructors. There also were several aviation mechanics from the Odessa school who joined us. Through the local government, I found a room to rent in the home of a widow who had a daughter in her mid-twenties. Although they were of noble birth, their income was limited and they were glad to have me as a roomer. It proved to be a happy arrangement for the three of us.

I discovered very shortly that in order to create a flying school one had to work very hard. First of all, the ditch had to be filled in and the electric wires and posts had to be relocated as far away as possible from the field. When I asked members of the local soviet for assistance, they assured me that they would help us as much as they could because they were delighted to have the school established in Alatyr. They suggested that we use war prisoners to fill the ditch and to do whatever else might be necessary. They promised to send us a few dozen prisoners the next morning and, sure enough, the men showed up at the barracks on schedule, ready for work. Most of them were Austrians, some spoke Russian, and after giving them instructions, I was pleasantly surprised to see how quickly and expertly they performed their tasks. By the end of the day, they had completed a considerable portion of the work, but to my frustration the following morning no one showed up for work.

Investigating the reason for their absence, I learned from a committee of the prisoners that they wanted better food or they would refuse to work! Such an ultimatum by war prisoners perturbed me because I knew very well that Germans shot and killed rebellious Russian prisoners of war for the smallest infractions of discipline. I replied to their demand in Russian, which was translated sentence by sentence into German, and I reminded them of the many Russians killed by Germans for daring to protest against any grievances whatsoever.

"You men must start to work immediately! Otherwise, I shall ask the local soviet to send a platoon of Red Army soldiers to enforce my order! I promise you," I stated firmly, "I'll speak to the soviet about the quantity of food you're getting, but remember this, food is scarce in Russia and it's hard to come by."

They discussed my offer briefly, then agreed to finish the work. They performed their jobs expertly and I kept my word regarding an increase in their food rations. Prisoner-carpenters converted the barracks into hangars for planes, quickly and skillfully; the electric wires were relocated and the landing field was now ready for a trial run. A single-seater Nieuport 17, a new fighter, was assembled, and I took it up for a test flight. After landing, I examined the operating condition of our new airfield. Everything looked excellent to me. I now fully expected to have a peaceful, pleasant life for many days to come.

My optimism proved to be short-lived. The fratricidal civil war between the Red and White armies had begun to engulf vast regions of Russia. Millions of people were dislodged and thrown into the all-consuming holocaust of civil war, and my dream of tranquility was for naught.

6. Escape to Kazan

At the time of Alexander Riaboff's defection to the White Army at Kazan, there was a major armed struggle underway, with the fate of the Bolshevik regime at stake. In May 1918, elements of the Czech Legion, an army of more than 40,000 former Czech and Slovak prisoners of war, seized partial control of the Trans-Siberian Railway. This bold move sparked a wider uprising against the Bolsheviks that included the Social Revolutionaries in Samara, Admiral A. V. Kolchak in Omsk, and General Anton Denikin in the south.

These were perilous times for the Bolsheviks and they spared no effort to defeat the assorted counterrevolutionaries outside Moscow. Leon Trotsky assumed personal command of the Red Army fighting at Sviyazhsk, just west of Kazan. The Bolsheviks made extensive use of aircraft in the Kazan campaign, mostly for reconnaissance missions, but on occasion for bombing enemy targets. Riaboff's Bolshevik commanders applied a harsh discipline on his air unit, a policy reflecting the extreme emergency faced by the Bolsheviks. Former tsarist military pilots were needed, but they were always under suspicion. Riaboff's account of this highly dangerous situation shows the role of political commissars, who were appointed by the Bolsheviks to assure discipline and political loyalty within the Red Army.

The Kazan campaign ended with a Bolshevik victory, but in the course of this conflict, Riaboff would make his way to the White side.

■　　■　　■　　■

THE 1917 REVOLUTION EVENTUALLY CAME to Kazan, at that time a large city, famous for its university, the alma mater of Lenin. In the early part of 1918, the Soviet Government had been established in the city, but in June of that year, a group of former army officers, assisted by other dissidents, revolted against the Soviet regime and took over the government. In ancient times, Kazan had been the capital of the Tartar Kingdom, which lasted until 1552, when Ivan the Terrible, after a long siege, took the fortress and returned the area to Russia.

These two military pilots and friends of Alexander Riaboff escaped to the White side on September 7, 1918. During the battle for Kazan they had been ordered to fly their Sopwith fighters from Alatyr to Sviyazhsk, the Red Army headquarters. They flew instead to a White-held airfield near Kazan.

The Bolsheviks were shocked by this sudden loss of an important city so close for Moscow and they decided to recover it. They ordered a large number of Red Army units with artillery and aircraft to be sent to Sviyazhsk, a small town on the right bank of the Volga just thirty miles west of Kazan. In view of this, our school received a wire from Moscow on August 10, 1918, ordering us to send all pilots and aircraft to Sviyazhsk immediately.

At the time of this crisis, I was the only operational pilot at the Alatyr school. My friend and colleague, Ensign Golov, had argued with Captain Chekhutov and departed for Moscow. When Chekhutov read the telegram to me, "send all pilots and aircraft to Sviyazhsk immediately," I realized that the honor of defending the Revolution would fall upon me alone, and I could hardly decline that honor.

As a student, I had fancied myself a liberal, generally speaking, but for the most part I had avoided political commitments. My energies were concentrated on my engineering courses and there was no time for studying political science, economics, or socialism. I was an ignoramus when it came to Communist ideology, and my judgment of the ruling Bolshevik party at that time was based solely on practical experiences. These encounters with the Reds had not been positive; many excesses and brutality were manifested in their acts. I must admit that I dreaded my new assignment, yet I had no choice but to fulfill it.

I asked to have my two best mechanics service my airplane. While I took off in my Nieuport 17, they were sent to Sviyazhsk by train to join me. After one flight, I was circling over Sviyazhsk looking for the airstrip. Next to the city's railroad station, which at the time was completely choked with passenger trains, I spotted several aircraft parked in an open field. I flew a wide circle, began my final approach, and descended on a straight line for the airfield. When my airplane reached the edge of the field, I noticed a fellow with a rifle aimed at me. He fired once, reloaded, and fired again and again. I had no choice but to touch down near the beginning of the strip, so I aimed my airplane at the man and continued my descent. As I approached him, he finally ran away and I landed safely. Several officials met me and escorted me to a passenger railroad car. There I met a dozen pilots and navigators, all former tsarist officers.

After some polite introductions, I found only one pilot, a Captain Kudlayenko, formerly of the Imperial Russian Air Force, whom I had known previously. Pilots, as a rule, are friendly with one another, but I sensed in this context that the usual camaraderie was absent. I felt an aura of tension and suspicion surrounding me as though I had been taken for a spy. In order to disperse this cloud of suspicion, I explained that I was an instructor in the newly established aviation school at Alatyr and that I had

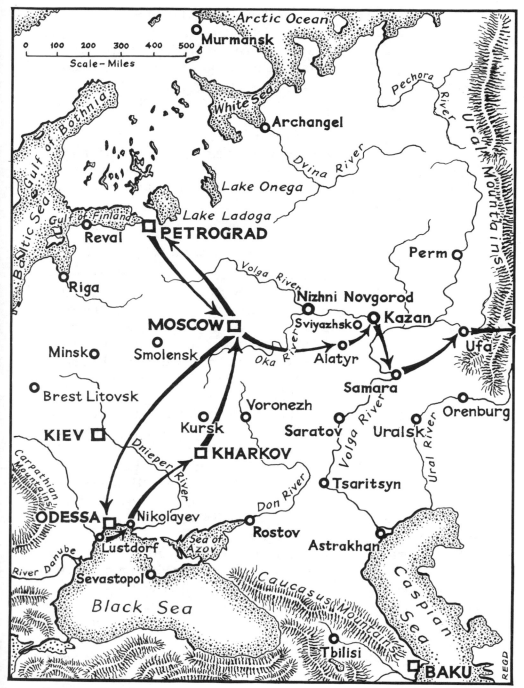

European Russia

This map shows Alexander Riaboff's movements during the revolutionary year of 1917, and subsequently during the early days of the civil war.

been ordered to join the Red Air Fleet at Sviyazhsk. They accepted my story with stony silence. In such a setting one had to be careful of one's choice of language. Often what one did not say or what one hinted at conveyed lasting impressions. Along these lines I made some remarks that suggested that I was not in Sviyazhsk necessarily as a volunteer. These efforts, it appeared, were to no avail. The cold reception had not been reversed.

After a few minutes the men began to leave the car. Kudlayenko and I were left alone. I was distressed over my apparent rejection by my colleagues. My friend Kudlayenko, noticing my state of mind, attempted to cheer me up: "Take my advice and don't talk politics or hint at your likes or dislikes—simply bide your time."

I followed Kudlayenko's advice for some time. But the temptation to break my silence and voice a request became too burdensome. One day when I approached our aviation commissar, K. V. Akashev,* I asked, cautiously: "Commissar Akashev, could I return to the aviation school in Alatyr? I would like to continue teaching there and I feel that they need me." Akashev had been sent to the area to organize Red air operations and was known for his toughmindedness: "No, Comrade Riaboff, you will remain here and you will fly reconnaissance missions without carrying bombs!" This statement appeared irreversible. And it was the kind of assignment for a pilot who has been classified as a counterrevolutionary!

Soon after this conversation, a new commissar was assigned to shepherd our air unit, so to speak. He was a fierce Communist and very proud of it. Others had treated us with respect and tolerance, but he raged and shouted at us. He was more threatening than Akashev.

In one of our conversations he asked me, "Have you read Nietzsche?" I confessed that I had not. "Nietzsche said, 'Help a feeble one to die.' Remember that!" he concluded with great emphasis. I felt that his remark was a sinister threat. His behavior had alienated our air unit in a significant way, not to speak of the fear it prompted.

In my opinion the Bolsheviks made a dreadful mistake by threatening and terrorizing everyone. They should have lectured us on communism and the teachings of Marx. They should have stated their plans for reform and the rebuilding of old Russia into a new, modern state. Instead of appealing to our minds, they ordered us to fight and die for an unclear ideal. We were young, intelligent, and liberal in our political outlook for the most part; and some of us might have accepted the Bolshevik's plans for government reorganization. To convert backward Russia to a Marxist

---

*One of the first commanders of the Red Air Fleet, the forerunner of the present-day Soviet Air Force.

During the civil war that followed the Bolshevik Revolution, the Red Air Fleet used the railroads to transport military aircraft to the most critical fronts. In this way they could obtain maximum use of their limited air power.

state was a formidable task and the new regime should have tried to attract the intelligentsia for this purpose by using an altogether different approach.

On my third day in Sviyazhsk, Commissar Akashev asked me about the number of machine guns and the amounts of ammunition, gasoline, and spare parts available at the Alatyr school. I told him the truth regarding the supplies there, and then I remembered a rumor that was being circulated at school that there were 100 machine guns on hand. Akashev then surprised me with the news that Chekhutov had been arrested for concealing government property and for sabotage.

I realized that if I had been in Alatyr, I would have been arrested along with Chekhutov as his assistant and accomplice! For the next two days I was not permitted to fly. One of my mechanics, whom I had no longer needed, had gone back to the school and had been arrested for some unknown reason. I fully expected to be arrested that night, but as time went on and no one apprehended me, I finally fell asleep, to awaken the following morning to a new and unpredictable day.

That next morning, while walking to the railway station at Sviyazhsk,

I met Jean Devoyot, a Russian of French descent and a former cadet pilot at the Odessa school. We were very happy to see one another and we discovered that we had much to talk about. As a matter of precaution, we walked to a nearby grove to continue our conversation. Seeing no sign of human life there, nor any evidence that anyone had been in that vicinity lately, we reminisced about our lives at Odessa, talked about old friends, and then turned our attention to the unpleasant circumstances of being under the Reds.

Jean gradually felt free enough with me to admit that he had become disenchanted with the Revolution and had decided to leave Russia and go to France to join the French Air Force. He told me he had come to Sviyazhsk to become allied with the White forces in Kazan, and as soon as his friend Neviazhsky assembled a reconnaissance plane at Alatyr, he would fly to Sviyazhsk, whereupon the two of them would fly over to the Whites at Kazan. Jean very generously suggested that I join them.

We walked slowly, completely engrossed in our conversation and totally oblivious to our surroundings, when we heard some faint noise behind us. What if someone had heard our conversation? For a moment, we froze in our tracks, then turned around slowly. There, standing behind us, was a poor, old peasant woman in rags, probably a beggar, who appeared to be quite ill. She couldn't possibly be a member of the dreaded Cheka, the Bolshevik secret police, "Thank God," I murmured to myself. We were both relieved—and were reminded again of the tenuous nature of our existence at Sviyazhsk.

Neviazhsky flew to Sviyazhsk four days later. He and Jean had worked out a plan whereby Neviazhsky would make a test flight in his aircraft, Jean would accompany him in a Nieuport, and they would both land in Kazan. The plan, however, went awry when a Bolshevik commissar unexpectedly appeared as a passenger for Neviazhsky's test flight, and a loyal Communist pilot was assigned to accompany them in a Nieuport fighter. This was an unmistakable sign to Neviazhsky that he was under suspicion, and he became very nervous.

Neviazhsky's aircraft was brought to the starting line, the commissar got in and ordered Neviazhsky to start the motor. In a few moments both were circling above the field. Suddenly, the engine stopped! Falling from the sky and out of control, Neviazhsky's airplane landed in a swamp, turned over and spilled its contents, including the badly scratched pilot and a commissar with a broken leg.

To the surprise of those who ran to the scene of the crash, a number of unexpected items belonging to Neviazhsky had fallen out. These were unusual items, the sort of cargo one does not take on a short test flight. Included were spare parts for the motor, a machine gun, and the pilot's personal belongings. One Bolshevik commissar became particularly interested in the collection Neviazhsky had found necessary to take on a "routine" flight. As a result, Neviazhsky was placed under guard in a hospital.

Jean and I had an uneasy feeling that we were also under suspicion and that our futures were more bleak than ever. We decided to leave Sviyazhsk at the very first opportunity. When Jean told Neviazhsky about our decision, he said, "Do whatever you find necessary, and leave me to my fate." That same day, August 19, 1918, we planned to go over to the White Army in Kazan. Jean was assigned to protect a reconnaissance scout and I was assigned in turn to protect the aircraft, flown by Kudlayenko, that carried a commissar. All the aircraft took off without difficulty.

I followed the airplane assigned to me until I saw my opportunity to turn mine toward Kazan. Once I had made this turn there was no going back; I had defected to the Whites. This was a fateful moment in my life. I flew directly toward Kazan, not knowing what sort of reception awaited me. When I reached the area of the Kazan airport, I made several circles, looking for Jean, but he was nowhere in sight. Several moments of painful indecision followed—should I remain with the Reds or should I take my chances with the Whites? I went down. As I descended toward the White airfield I spotted Jean's aircraft landing at the same time! This happy circumstance renewed my spirits greatly. A short distance away from the landing strip, a group of men stood beside a building observing us. They appeared to be very excited, and the moment I landed they surrounded my airplane and greeted me enthusiastically as though I were a hero. They told me that just ten minutes before my arrival another machine, piloted by Yefrimov, had landed. No wonder they were excited: three defectors from the Red Air Fleet in one day was a bit unusual.

We were taken to the headquarters of the White general staff in Kazan and were interrogated regarding the strength and disposition of the Red air units. Being frontline pilots, we were unable to supply our hosts with such information on other units. That same evening, we were feted at an official banquet in the residence of a wealthy Kazan merchant. In attendance were members of the general staff and the provisional government of Kazan.

The following day, a local newspaper reported that three Red aircraft had been downed by the White Army in Kazan. We had requested this falsification of the circumstances surrounding our arrival in White territory in order to avoid possible reprisals against our families. Our attempted coverup was in vain, however, because the Red Army had known about our defection on the very day it occurred. The Bolshevik political commissars were furious and immediately threatened reprisals against our families, but the remaining pilots in the Red camp protested so strenuously that the commissars withdrew their threats.

Sophia Dmitrievna Nikitina

A lexander Riaboff met his wife, Sophia Dmitrievna Nikitina, while fighting the Bolsheviks in western Siberia in 1919. Sonya, as she was known, the youngest of four children, was born in the city of Pskov, in 1898. Her father, Dmitriy Nikitin, worked for the railroad and died in an accident when Sonya was only five months old. This tragic circumstance left Nikitin's widow, who never remarried, with the formidable task of providing for her four children. As a child, Sonya Nikitina dreamed of becoming a doctor. The advent of World War I in 1914, when she was sixteen, prompted her to persuade her mother to permit her to enroll for training as a nurse with the Russian Red

Cross. The war afforded the young and innocent girl a hurried pathway to a career of sorts in medicine, even as it set into motion events that would alter her life dramatically.

With minimal training, she found herself attached to the Russian Second Army at Novogrigoriyev, near Poland, beginning in October 1914. Here she served in a poorly equipped field hospital, with few doctors, nurses, or medical supplies. For the young nurse, who in prewar years had led a rather sheltered life, the carnage of World War I made a deep and lifelong impression. Working at a frontline hospital, she was surrounded by thousands of wounded and dying soldiers. The staff struggled in vain to meet the crisis, attending as best they could to the needs of countless wounded lying on the floors and hallways of the hospital. She remained at that post for about one year.

Soon the war reached Sonya Nikitina directly as the Germans launched an offensive against the Russian Army in 1915. The hospital itself came under fire, being bombed by the Germans on several occasions. This experience severely effected the young nurse's health, forcing the Red Cross authorities to allow her to return home briefly to recover from shell shock.

After her furlough, she returned to the front, just in time to be captured by the advancing German Army. Sonya Nikitina, along with other members of the hospital staff, were removed to Germany as prisoners of war. This period of captivity brought abuse and ill-treatment at the hands of her captors. In late 1916, the Germans released her from prison, and upon her return to Russia, she received a number of medals and awards for her contribution to the war effort. But these medals did not erase the trauma of the war or the rigors of her captivity. She emerged from this experience with a keen dislike of Germany for its brutality and a growing bitterness toward the tsarist government for its indifference to the suffering of Russian soldiers.

During the war Sonya Nikitina's family had moved to Samara (Kuibyshev). She spent most of the revolutionary year of 1917 in Samara, recovering her health and reestablishing contacts with her family. Once the violence of the Revolution had engulfed Samara, she and her brother Vassiliy decided to move east, toward Siberia, to find work. It was during this period in Siberia that she met and married Alexander Riaboff. While the young couple were on the move across Siberia, Sonya nursed Riaboff back to health when he contracted typhus. Her brother Vassiliy, however, died of that disease during this same period.

Sonya escaped to China with her husband in 1920. She had one daughter, Helen, who was born in 1922. The following year the Riaboffs emigrated to the United States. Sonya died in San Francisco in March 1945.

The Nikitin family, except for Sonya, remained in Russia after the Revolution. Despite difficult times in the immediate post-revolutionary period, the family managed to adjust to life under the new regime. Sonya's mother died in 1939. Her surviving brother (Dmitriy) and her sister (Yelena) died in 1964 and 1968, respectively.

Many years later, when I returned to Moscow in 1960 for a visit with my sisters, they told me that in the autumn of 1918, two agents of the dreaded Cheka had visited my family to inquire as to my whereabouts. During their visit they searched my room. This was the only time, fortunately, that my family was threatened by the Bolsheviks on account of my defection.

Although the civil war continued to be fought in several areas of Russia, air battles between the Reds and the Whites were very rare. Perhaps this could be explained by the fact that pilots felt they belonged to a fraternity and, therefore, preferred to help one another rather than "fighting to the finish."

A few days after landing my airplane in Kazan, I received a dreadful shock. On my way to the airfield I stopped by headquarters to confer with those in charge of the White air operations. Everyone there seemed to be in a depressed state of mind and I was curious to learn the reason for the somber mood. I approached Captain Boreiko, former commander of the Gatchina Military Flying School, and now in charge at Kazan, and asked why everyone was so somber? "The situation is critical," he replied despondently. "Our troops are being hard pressed by the Reds west of Kazan. If I could send two or more pilots to that area to strafe them from the air, we might be able to thwart their efforts somewhat."

This crisis was my opportunity to offer my services as a pilot with a reliable fighter equipped with a machine gun. "I'll be glad to go anywhere you suggest and follow whatever orders you give me," I offered. "My good man, if only you could! Your aircraft was wrecked the day after you arrived by some greenhorn who tried to fly it!" I was crushed with disappointment. My wonderful Nieuport 17 was beyond repair, I was told, leaving me grounded.

My friend Jean had left Kazan for Samara on his way to Paris. Yefrimov had made a reconnaissance flight for the White Army and told of a fantastic series of aerial dog fights, but everyone suspected that the boastful Yefrimov had fabricated the story. These exaggerations not only put him in a bad light and led to his grounding, but inadvertently affected my life. Because there was no airplane available for me to fly, and Yefrimov had been grounded, the two of us decided to try our luck in Samara, a city located about 200 miles downstream on the Volga

Our departure for Samara came at a time when the White military fortunes had waned. It is of interest to note that the Red forces bringing pressure upon the Whites in Kazan were under the command of Trotsky, who at that time served as Lenin's right-hand man and organizer of the Red Army. This particular area of conflict was one of the most important arenas of conflict for the Reds. They were determined to recapture it and push the

Whites eastward to the Ural Mountains. Since the Reds had cut the steamship communications between Kazan and Samara, approximately twenty miles down from Kazan, we were forced to walk through Tartar* settlements and villages part of the way and then find horse-driven transportation for the balance of the journey. At a large Tartar settlement on the Volga, we boarded a steamer headed for Samara. At no time along the way did we encounter any Red Army units, partisan bands, or Bolshevik sympathizers in Tartar territory.

The Tartars proved to be sympathetic toward our cause and they expressed a keen dislike for the new Bolshevik regime. They often struck up conversations with us, offering us food and lodging as night approached. It was obvious to them that we were traveling without provisions of any kind. After all, we had been unable to take our gear with us at the time we switched sides. We finally arrived in Samara, found the aviation school, and offered our services to the local White Army commander as military pilots. When we arrived at the airfield, there were no available aircraft for us to fly, so we were assigned to the aviation school as reserve officers. Once again we found ourselves unoccupied, bored, and frustrated. Pilots without airplanes feel as useless as Cossacks without horses.

On September 7, 1918, my friends Captain Kudlayenko and Lieutenant Neviazhsky landed in a Sopwith aircraft at the Samara aviation school. They had been ordered by their Red commander to fly from Alatyr to Sviyazhsk, but on their way they changed their course to the direction of Samara. Two more defectors were eager to join the Whites and to divulge the latest aviation news from the Red side.

We learned that the aviation commissar who had witnessed Neviazhsky's crash and incriminating cargo at Sviyazhsk, when he had attempted to fly to Kazan, had intended to shoot Neviazhsky then and there, but the pilot, being a charming and personable chap, had talked him out of it. We were also told that the Alatyr school was barely functioning, that the Moscow Aviation School's commander had been arrested, and that several pilots had been shot in Moscow after having been accused of belonging to a secret White Army organization. As there was nothing for Kudlayenko and Neviazhsky to do in Samara, they proceeded on to Siberia to join the Whites there.

With each passing day, the military situation for the Whites at Kazan and Samara worsened, finally forcing the White Army to order the evacuation of Kazan for Ekaterinburg along the connecting railroad and from Samara eastward along the Trans-Siberian Railroad. The Whites left Samara on September 25, 1918, with a plan to reestablish the aviation school

---

*The Tartars were the descendants of the Mongol invaders of the thirteenth century.

some place west of the Ural Mountains. Instead, the White Army was forced to retreat well into the Ural Mountains under the constant pressure of the Red Army. Because of this military pressure, a Lieutenant Dudka and I were ordered to proceed to the town of Byisk, which was located south of the beautiful city of Novonikolayevsk (present-day Novosibirsk). Our orders were to seek out a site there for a new aviation school.

We found Byisk to be a delightful town with a moderate climate, situated near the snow-capped Altai Mountains. It was connected by railroad to Novonikolayevsk. After exploring the area, we concluded that it was not a suitable place to establish the school. There was no large level space for an airfield, nor were there any buildings to house the school. Within a short time we received a telegram ordering us to go to Kurgan where another school was being set up about two miles from the town.

Upon arriving in Kurgan, I was pleased to find my good friend Rozhdestvensky at the aviation school working as a flight instructor. He had found a room to rent, large enough to accommodate two, so he offered to share it with me. Housing was difficult to come by and I felt fortunate to have solved that problem so soon. It was also fortunate that my friend left the room early in the morning, returned late at night, six days a week, thereby making the arrangement ideal for both of us.

Many of us pilots were forced to remain idle. I tried desperately to find ways to fly and I was willing to go anywhere in order to fly. One day, while I was talking with a Captain Larshin from the Ural Cossack Army who was then engaged in fighting the Reds in the southwest Ural Mountain area from the city of Uralsk, he asked, "Why don't you join the Ural Cossack Army? We've captured several airplanes from the Reds, but we have no pilots to fly them."

When I told my friend Dudka and one of the school's instructors, Galetsky, about Larshin's offer, they agreed that it would be a good idea for us to try our luck in Uralsk. Our commander gave us permission to become pilots of the Ninth Aviation Squadron in Uralsk; the only problem confronting us was—how to get to Uralsk alive! Captain Larshin had a plan for us to proceed by railroad eastward to Chelyabinsk, a town in the Ural Mountains, then again by railroad going south to Kartaly, 100 miles away at the end of the line. From Kartaly we would procure horses and ride 400 miles over mountain roads to the city of Orenburg and then another 300 miles on the same types of roads to Uralsk. It would be an arduous trek, but at Uralsk there was a real prospect of flying again.

We then met with Larshin. "I know those roads like the back of my hand and I assure you we won't have any trouble in arriving at our destination," Larshin said emphatically. "Aren't we liable to meet Red bands on the way?" I asked. "You can be sure, none at all." he replied.

Alexander Riaboff at the time he escaped to join the Whites in the civil war.

With his positive assurance that the trip was feasible, we set the date for our departure. In preparing for our risky journey, I thought it necessary to make several trips to the Siberian White Army's headquarters in Omsk to inquire about the existing military situation and to check on the available transportation. The more I thought about a rugged journey over mountain roads on horseback to reach Uralsk, the more I had doubts about the whole project.

Even as these events were unfolding, my life had taken another important turn. On one of my trips to Omsk, while waiting for a train at the station, my eyes singled out a girl standing in the midst of a large crowd of people. I thought she smiled at me, but her smile could have been directed at someone standing directly behind me. Well, I thought, I'll take a chance and find out if I'm the lucky one. All she can do is turn away from me.

I walked over to her, looked directly into her large, blue, friendly eyes and said, "May I introduce myself?" I'm Ensign Riaboff—Alexander Vasilievitch. Waiting endlessly in railroad stations for trains that seldom arrive is an awful bore, don't you agree?"

"Hello, I'm Sophia Dimitriyevna. Yes, I agree with you. Are you headed east or west?"

"I'm returning to my aviation school in Kurgan, and you?"

"My brothers live near Kurgan and I am living with them at the present time. So, we are both traveling west."

Suddenly, an announcement blared over the crowded railway platform: "RAILROAD TRAFFIC IS DISCONTINUED FOR THE NEXT TWENTY-FOUR HOURS BY GOVERNMENT ORDER." We then parted and went our separate ways until the following day when the time arrived to return to the station. Sure enough, there she was, and this time I knew her smile was meant for me.

I searched for a train going west and found one with a boxcar containing passenger seats and two vacant places. I boosted my newly found friend aboard and soon the two of us were on our way. As soon as we were settled in our seats I said, "Well, now, Sonya Nikitina, tell me all about yourself. By the way, my relatives call me Sasha, and I presume you're Sonya. Let's drop the formalities. In precarious times like these we can't waste a moment and I do want to know all about you."

She told me she was born in Pskov in 1898, the youngest of four—two boys and two girls. Her father had died before she had a chance to really know him. After his death, her brothers supported the family, working as mechanics. She told me about the war years: "I attended school until the beginning of the war in 1914, when I decided I wanted to be a nurse. Most of the boys I knew in Pskov had been killed by the Germans and I began to feel a tremendous urge to do something to help in the war effort. When I learned about a special school for nurses' training, I applied, was accepted, and completed the course in a few months, and was immediately sent to the front line of combat." She stopped abruptly and began to cry.

It was then that I realized that the girl must have lived through some horrible trauma because she was unable to continue until her sobbing subsided. I felt deep compassion for her. Attempting to console her, I said, "Please don't go on if your memories make you suffer so much." But she seemed to want to continue anyway: "I have seen all of the horrors of war. I have actually waded in human blood. My nerves are ruined because of gas poisoning, and I have been wounded by an exploding shell. The Germans occupied our military hospital when they defeated us and held us prisoner for a while—what beasts they were! I have been decorated for bravery—decorated for what? Who wouldn't have taken care of our poor, maimed boys, begging me to help them in their misery as they lay suffering and dying! War is the most disgusting, obscene display of man's cruelty to man in all of this dreadful existence. I have wanted to die—so that I'd forget the horrors I have lived through."

I put my arm around her shoulders, but failed to find the right words to console her. My experiences could not be compared with hers. I had been spared the worst, and I wondered then if her battle wounds would ever heal. We both fell asleep for a while, our heads resting on each others'. When we arrived in Kurgan, I said goodbye to her, knowing it was not a final farewell. She continued on to the small town of Chumliak where she lived with her brother Vassiliy. We had promised to write to one another as often as possible, which we did quite consistently. At my first opportunity I went to Chumliak to visit her.

7. On the Move

After joining the Whites at Kazan in 1918, Riaboff found himself on the move again when the Bolsheviks captured this historic city on the Volga. By the spring of 1919, the Whites in western Siberia, then under the leadership of Admiral A. V. Kolchak, were ready to launch another major offensive. This campaign enjoyed some initial successes as Kolchak advanced from Perm toward Vyatka (Kirov). Other White armies advanced from Archangel in the north while a huge force of 300,000 men under General Denikin invaded from the south, capturing Tsaritsyn (later Stalingrad and still later Volgograd). Outside Petrograd, General Nicholas Yudenich made an abortive attempt to dislodge the Bolsheviks from the old imperial capital.

The Bolsheviks launched a series of powerful counteroffensives that quickly shattered the hard won White victories. By November 1919, the Red Army captured Omsk, Kolchak's capital. In the wake of this Red drive there was considerable confusion along the Trans-Siberian Railway as the Whites retreated and frantically attempted to stabilize the front. Kolchak himself was betrayed to the Bolsheviks and was executed. As it turned out, Allied intervention in 1919 was not sufficient to influence the course of events favorably for the Whites.

For Alexander Riaboff these were extremely difficult times. While there were occasional tactical victories by the Whites over their Red pursuers, Riaboff found 1919 to be a disastrous year.

■   ■   ■   ■

A FEW DAYS BEFORE OUR PLANNED DEPARTURE for Uralsk, I became ill with a severe cold and was unable to go with my companions. Captain Larshin said, "If it's agreeable with you, Galetsky and I will proceed to Chelyabinsk and wait for you and Lieutenant Dudka." Several days later Dudka and I arrived in Chelyabinsk, looked for our friends at the appointed place where we had planned to meet, and were told that Galetsky and Larshin had already departed for Uralsk. Since we could not make the journey without Larshin's assistance, we returned to Kurgan.

We reported to our commander, who then gave us some promising information: "An automobile column with military supplies for Uralsk has

Alexander Riaboff in a bearskin coat next to his Nieuport 17 fighter.

left Omsk and will pass through Kurgan in a few days. You and Dudka can join the column, but the chances of its ever reaching Uralsk are very slim." When I contemplated the problems involved, I had to agree with him.

My illness persisted. Just before New Year's day of 1919, I suddenly took a turn for the worse, forcing me to seek the help of a physician. While he was questioning me about my condition I realized I was having difficulty understanding him or his questions. My disorientation to his preliminary examination irritated him to the extent that he chose not to examine me further. "You're O.K., Riaboff. Go home and get some sleep." His diagnosis could not have been more inaccurate. With great difficulty, I managed to return to my room, and for the next few weeks I struggled between life and death with typhus.

My landlady discovered my condition the next day, took pity on me, and treated me as a helpless child—which I was. I rapidly became worse, lapsed into unconsciousness, and was delirious. Constantly changing colored lights emanating from a central core outward, thereafter losing their intensity and disappearing, somewhat like a kaleidoscope, played before my eyes.

One episode in particular stands out in my memory as though it were a recent occurrence. As I lay on my left side with my eyes open, staring into emptiness, a bright, familiar object came suddenly into view. It took some time for my eyes to focus and for my weakened brain to accept what I was seeing. Finally, I recognized that "familiar object" to be my own head! It protruded from under my bed and was staring at me quite impersonally.

"How can my head by there?" I asked myself. "It should be here," and I made a mental effort to return it to its proper place, but in vain. I became exhausted from my attempt to accomplish my impossible feat, until once again unconscious oblivion engulfed me. I recall the onset of the disease and my gradual recovery, but very little in between except for my misplaced head incident.

As soon as Sonya learned of my illness she moved to Kurgan to take care of me. She and my landlady worked together to pull me through and I was grateful that neither they nor my roommate became infected with the dreaded disease.

When February came, I renewed my efforts to go to Uralsk. After numerous delays, the automobile column trip to Uralsk was finally canceled because Orenburg and Uralsk had been taken by the Reds. Our dream of reaching Uralsk had evaporated. At that time Dudka and I were in Troitsk, which became the center for the retreating Orenburg Cossack White Army. Dudka returned to Kurgan and I was ordered to serve as an aviation staff officer for Colonel Kaplin, chief of aviation for the Orenburg Cossack White Army. "Aviation" proved to be a misnomer, for there were

Alexander Riaboff, sixth from the left, attends a Russian Orthodox divine liturgy in an open field with members of his squadron.

no aircraft or air personnel. My promotion had not changed my status as a pilot without an airplane and with nothing to do!

About this time, the bad news reached me that Captain Larshin and Galetsky had been caught by a Red band and killed while they were on their way to Uralsk. To think that I would have suffered the same fate had I been able to travel with them! Ironically, typhus had saved my life. This same disease was rampant at the time, and had killed thousands as it swept through Russia. Miraculously, I had recovered from it with no ill effects!

Three months passed in Colonel Kaplin's office, while I did nothing. Finally, this inactivity was enough for me and I decided to take action. I wrote a letter to my friend on the aviation staff of the White Army in Omsk, requesting his assistance in getting me an appointment as a pilot in one of the active squadrons fighting against the Reds. On May 8, 1919, my chief received a telegram from Omsk stating that I had been appointed to serve as a pilot in the Tenth Aviation Squadron. Colonel Kaplin asked me to stay on with him, but I declined because I chose to take part in a real war for which I had been trained as a combat pilot.

On my way to join the Tenth Squadron, I decided to visit Sonya, with whom I had been corresponding steadily. We spent several days together and gradually became aware of the fact that we were deeply in love with one another and wished to spend the rest of our lives together. We chose the city of Ufa in which to be married, and on May 18, 1919, our wedding ceremony was performed in a Russian Orthodox church there. After a three-day honeymoon, we left together for my newly assigned duty with the Tenth Squadron.

Captain Astafiyev, the commander of the Tenth Squadron, proved to be a very cordial gentleman. He entrusted me with his Nieuport 17 fighter,

The White air units mobilized military aircraft wherever possible. Riaboff's squadron, shown in the early days of the civil war in Siberia, contained a variety of Nieuport, Sopwith, and Morane aircraft.

which I flew successfully on my trial flight and thereafter flew regularly. It so happened that the Reds had no aviation in that area and our flights were confined strictly to reconnaissance. There were seven pilots in our squadron and each flew his own airplane. Our mechanics were not only good, but they were reliable as well. We operated as one friendly family, thus creating a comfortable, secure atmosphere. Unfortunately, we were forced to change our location often and sometimes suddenly, depending on the situation of the war. Airports and landing strips were nonexistent. In selecting a new location, we would fly over a level meadow several times to determine its suitability for a landing field.

Before we had to move on to a new location, a Red Army pilot, Lieutenant Shimaniyel, landed on our airstrip and brought me certain information from the Reds. "I saw your name on a Red Army order as a defector. The order also stated, 'In the future, defectors' families will be arrested and shot!'" Apparently the order had not impressed the recent defectors, because that same day another defector pilot, Ensign Muratov, landed his plane on the Thirty-third Squadron's strip adjoining ours.

On one of my reconnaissance flights, I carried out an order to inspect carefully a certain forest area north of the Trans-Siberian Railroad, to learn whether or not a Red Army unit was hiding there. I took off before sunrise, found the forested area, and descended until I was just above the tree tops. The forest was congested with Red Army infantrymen. Some were sleeping, some lying or standing under the trees, and I was close enough to see their faces. They actually smiled at me, evidently mistaking me for a Red pilot. My arrival had neither caused a commotion nor had anyone taken a shot at me.

I flew on further to take a look at the nearby village where everyone seemed to be asleep, and I saw a large number of horses and vehicles indicating the presence of a considerable armed force. My fighter and I returned to the base unscathed and my subsequent report was appreciated by the staff.

In October of 1919, I was ordered to go on another routine reconnaissance flight. At that time I was flying a two-seater Sopwith with a machine gun attached to the back seat, thereby permitting the navigator to have a full-circle firing range. My regular navigator was absent so I was obliged to take a Lieutenant Kashin as a replacement. It so happened that Kashin had never been in the air before and had no idea of what might be expected of him in case of an emergency.

Periodically, I looked back at him. He was smiling at me, indicating that he was enjoying his first flight to the fullest, when out of nowhere an airplane appeared. It was flying about a thousand feet to my left, going in the opposite direction and being pursued by another plane at a dangerously close distance. In the lead was a Sopwith going unusually fast while the pursuing aircraft, a SPAD, displayed a red, five-pointed star on its fuselage. A Red fighter—my very first encounter with one since I had entered the civil war!

I knew that the Sopwith belonged to Volkovoinov, one of our pilots who had preceded me on a reconnaissance flight. A SPAD was attacking him, and Volkovoinov was in full flight, heading for our airfield in an attempt to save himself and his navigator. The SPAD, a much faster, more maneuverable plane, had the advantage over the Sopwith and I was afraid that my friends' chances of escape were slim. I reacted promptly to the situation, turned my airplane sharply to the left, and headed straight for the

Gasoline lorries used by Riaboff's squadron in the Siberian campaign against the Bolsheviks. Spare parts and tools were kept in the adjacent tents. Chronic shortages in fuel and spare parts made air operations difficult even with aid from the Western powers.

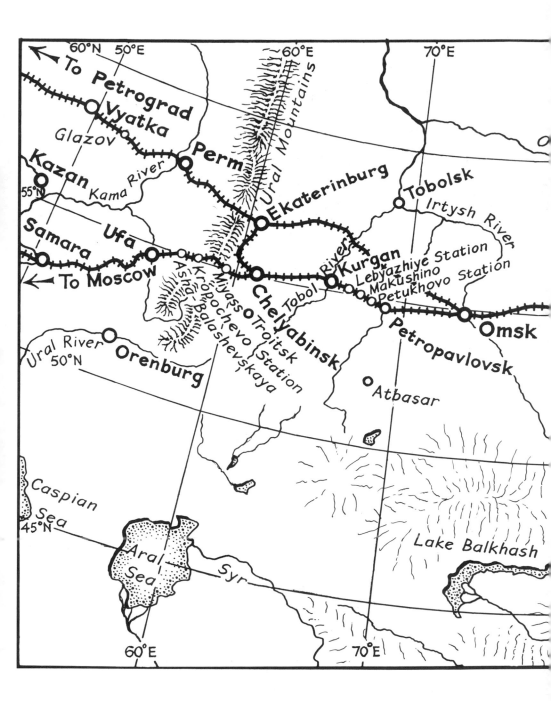

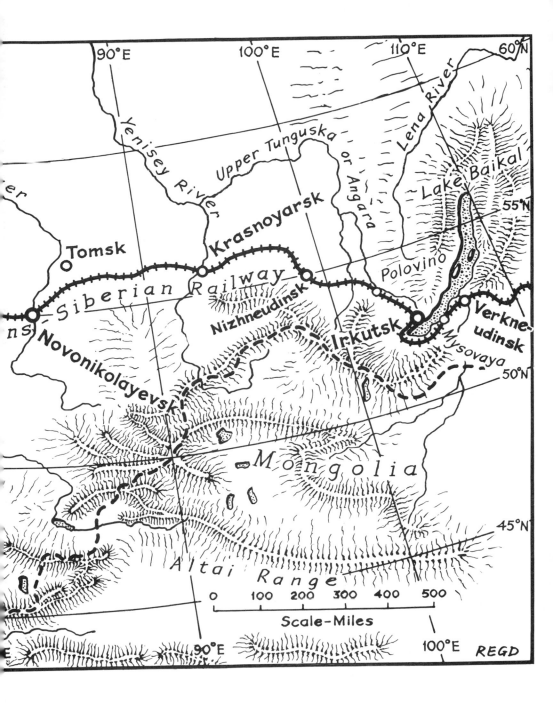

The Whites under Kolchak challenged the Bolsheviks in Western Siberia in 1918–1919. Alexander Riaboff defected to the anti-Bolshevik side at Kazan. Later he retreated eastward with the White forces along the route of the Trans-Siberian Railway.

SPAD. The Red pilot spotted my maneuver and abruptly turned his fighter, going straight for me. Now was the time for my navigator to use our machine gun. I looked back at him, gesticulated wildly, my left hand pointing directly at the SPAD, but he only smiled benignly back at me.

From where he sat, Kashin was unable to see the SPAD's impending attack. Vocal communication was impossible, my gesturing was in vain, the two planes were rapidly approaching one another at the same elevation and I expected a short burst of machine gun fire to come from the SPAD, but there was none. We passed each other at a distance of seventy-five feet and I could see that he was about my age and type. I turned sharply to the left, came up close behind him in a perfect position to fire, but my gunner was of no use to me whatsoever, and within a few seconds, the SPAD— traveling much faster than I—was completely out of my range of fire.

When he disappeared, I glanced back again at my navigator and there he sat with the same happy, complacent smile on his face, unaware of our miraculous escape. Taking everything into consideration, I decided to give up my chase of the SPAD and return to the airfield.

Upon landing, I was told that Volkovoinov's airplane had been attacked unexpectedly by the SPAD from the rear and his gas tank had been shot through, and for a moment the contused pilot had lost consciousness. His navigator, seeing gasoline pouring from the gas tank, attempted to stop the flow of the gas instead of returning machine gun fire. As luck would have it, I came to their rescue just at the right moment, but it had been a close call for everyone.

At the first opportunity, I questioned Kashin, my so-called gunner, "Where were you when I needed you so desperately?"

"I confess," he replied, "I did not see the other airplanes. I was not aware of your maneuvering the airplane for any specific purpose, and I understood your gestures to mean 'take a look at the interesting scenery below.' "

Later that day, Volkovoinov and his navigator thanked me for having saved their lives. All I could say was, "I think our guardian angel must have been with us. I will admit, however, I came along just at the right time too!"

When I told Sonya what had happened during the day she expressed alarm at my close call, but had to laugh at my frustration when I described the episode. "I kept pointing forward trying to signal Kashin to look for the Red fighter, but he just shook his head affirmatively with that silly grin on his face. I was really desperate and could do nothing about it." I began to see the humor in the situation, too, and we both laughed until we cried.

"Between you and me, I'm glad my gunner didn't get my message. In my opinion, it's just possible Russian pilots avoid the opportunity to kill

This hangar contains four Nieuport 17s and a Sopwith 1 1/2 Strutter undergoing repairs at a Siberian airfield used by Alexander Riaboff's squadron during the civil war. Note the Il'ya Muromets insignia at center.

Alexander Riaboff at the controls of a Sopwith 1 1/2 Strutter at the start of a mission against the Red Army. The observer takes a bomb and propaganda leaflets to drop on the enemy.

A squadron party. Muromtsev, the squadron commander, held the party on the occasion of a visit by his girl friend (center). Riaboff is seated at far right.

Captain Muromtsev takes command of the White Tenth Air Squadron.

each other. Civil war is a nasty business." I had concluded by this time that war was not for me, and under no circumstances would I ever choose a military career.

Captain Astafiyev, our former commander, had been appointed to serve on the aviation staff in Omsk. He was replaced by Captain Muromtsev, an excellent pilot and a courageous man who brought his own navigator, Lieutenant Voshchilio, with him. The two men were very good friends and equally brave, seemingly without fear. They also had a common weakness, which more or less set the tone for the entire Tenth Squadron, except for myself. Every night our commander led the group of officers in a drinking bout. It became a routine contest to see who could consume the most vodka. I had never drunk hard liquor and I had no intention of joining their contest.

Eventually, their excessive drinking became intolerable to me and I requested a transfer to another squadron. My transfer arrived very soon, ordering me to join the Fifteenth Aviation Squadron in Atbasar, a town located 250 miles south of the Trans-Siberian Railroad.

Before our departure, we went to Omsk in order to meet my new commander, Colonel Kompaniyetsev, who informed me that I was to remain in Omsk until further notice because the Fifteenth Squadron was coming to Omsk to be refurbished. The colonel revealed various bits of information regarding the whereabouts of pilots we had known. "It may interest you to know that Captain Muromtsev and his good friend Voshchilio failed to return from a reconnaissance flight and their fate is unknown." I was always saddened to hear about the loss of old friends, but in times of war it is to be expected.

Sonya and I found living quarters in Omsk, and we fully expected to remain there until the spring of 1920, but military events in Siberia quickly shattered this hope. The Reds had coordinated their forces on all fronts, launched a powerful offensive, and now threatened Omsk. A White government order arrived with instructions to evacuate the city and head eastward. The aviation staff was ordered to move its shops, personnel, and wives onto two specified trains. One car was a deluxe passenger type with a dining car and boxcars to carry shops, equipment, and supplies. The other was made up of boxcars only; some for housing personnel and some for shops and supplies. As I was to remain in Omsk with the staff, Sonya was to go east to Irkutsk or to the train's destination. Departure was scheduled for November 2, 1919.

Sonya and I went to a station where the trains waited on spur tracks to be dispatched. We brought all of our possessions, consisting of clothes and a few books, with us, but we discovered, to our great disappointment, that because I had failed to inform the authorities of Sonya's leaving, there was no place reserved for her on the deluxe train. However, we were told that some people would leave the train at Novonikolayevsk, located 400 miles east of Omsk, at which time she could transfer to the deluxe train.

As we stood in front of the passenger car waiting for the departure signal, Colonel Kompaniyetsev and his staff passed by. After exchanging salutes, he approached me and said, "What are you doing at the station?"

"I'm saying good-bye to my wife. She's leaving for the east," I replied.

"Aren't you going to accompany her?"

"No sir," I responded, wondering why he asked me such a question. He turned to one of his staff officers and snapped, "Do we need him?" and didn't wait for an answer, but continued, "You had better go, too, and take care of the passengers. Furthermore, you may try to find a suitable airfield with shops, if possible, in Irkutsk or east of the city for our squadron."

"Thank you sir, I'll do as you say, but may I please return to my room to get my clothes? I understand that the train will leave in a few minutes, so what do you suggest I should do?"

"Don't worry. We'll hold the trains until you return, but make it as fast as you can."

I was back in an hour and we were off. Our train followed the deluxe one as closely as possible, according to Captain Belokurov's instructions. He was in charge of both trains and had told Captain Grishin, in command of our train, to see that it never became separated from the first train. After settling ourselves in the boxcar, I said to Sonya, "Who could have imagined that the two of us would be heading east together?" "I guess we are just lucky," she replied, "And I hope our good luck continues."

A partition divided the front portion of the car into a compartment occupied by a middle-aged couple and their grown daughter. Evidently the man was an important government official, having received the more deluxe accommodation. A wood-and-coal-burning stove stood in the center of the car, and at the rear was a pair of bunk beds on either side wall. We were assigned the bunks on one side and two girl clerks had the other side. With a sheet hung between, we were allowed some privacy. There was ample space in which to move about, everyone seemed amiable, and I predicted we would have a pleasant journey.

Both trains arrived at the main Omsk station where numerous other trains were waiting on adjacent spur tracks for signals to proceed eastward. To all appearances, I figured we would have to wait at least one week before the trains could be dispatched, but to everyone's surprise, we were ordered to proceed after a short delay. We reached Novonikolayevsk, making a few short stops en route, and there we were shocked to find the station completely glutted with railroad cars.

The stationmaster informed me that we would have to wait for several days before he could move our train. With nothing better to do, Sonya and I decided to take a walk around the city. We had taken only a few steps when I heard, "Ensign Riaboff, come on back, our train received a permit to proceed at once." I turned around and saw one of our aviation mechanics excitedly waving us back to the station. Within five minutes, our train was moving away from a colossal mess of a hundred or more trains, including the deluxe one that had formerly preceded ours.

"Try to explain this miracle!" I exclaimed as Sonya and I watched our exodus from a labyrinth of tracks and trains. Crowds of people were milling about the station in utter confusion; some had been waiting for weeks to move on, and the fact that our train was well on its way to Irkutsk was nothing short of a miracle!

Later, I had an opportunity to ask Captain Grishin how he had managed to extricate our train and push it ahead of the others. "We had the good fortune to be carrying a considerable number of supplies, such as alcohol, flour, butter, and other food, all very difficult to procure in these chaotic times. I offered to donate a portion of our precious cargo to the stationmaster with his promise to put our train on the proper track and send us on our way in a hurry. I kept my part of the bargain and he kept his. A little bribery did the trick."

I could not help but admire the man's resourcefulness when confronted with an almost insurmountable problem. Finally, we reached Irkutsk, traveling a distance of about 1,000 miles, without difficulty or delay. From there we were ordered to continue on to Ulan-Ude, a town located approximately 300 miles east of Irkutsk, to look for a suitable place for an aviation field and repair shops.

Although railroad traffic east of Irkutsk was normal, our train was stopped and placed on a side track at a small isolated station. Because the weather was very cold and it was snowing constantly, there was absolutely nothing to do except stay inside our boxcar in an attempt to keep warm and to watch the trains carrying released war prisoners on their way home. For no apparent reason one night, a train transporting some of them stopped, took our engine, and continued eastward leaving us completely stranded!

Captain Grishin took immediate action to rectify the situation by organizing an expedition to carry a large supply of spirits and food to the first large station to the east of ours. He and his aides boarded the first Czechoslovakian prisoner train heading east, and within two days' time they were back with a locomotive. After it was connected to our train, we traveled nonstop to Ulan-Ude, remained there two days, and were ordered to proceed another 400 miles east to the city of Chita. By that time, it was the middle of January 1920—far away from civil war and the dreaded Red Army.

Captain Grishin's ingenuity and persistence saved our lives, but another guiding Force must have arranged our boxcar accommodations. All of our worldly possessions had remained on the desired deluxe train, as our intention had been to transfer to the better train at the first opportunity. That opportunity never arrived, nor did the deluxe train for that matter; it had been captured by the Reds a few miles east of Novonikolayevsk!

After my departure from the First Air Squadron in September 1919, news of events in Siberia concerning the White Army had become progressively worse. At that time, the front line was about 400 miles west of Omsk and seemed to be more or less stationary. Omsk appeared to be a safe distance from the Red Army, although the Reds had concentrated a large force northwest of the city, thereby forcing the Whites, operating along the Trans-Siberian Railroad, to retreat.

By November 18, 1919, Omsk was taken by the Reds, a conquest the Whites had not anticipated. The Red drive eastward had been rapid and decisive. Panic ensued when the White evacuees found themselves caught in the melee of stalled trains lined up along the railroad for miles and miles. There was no coal to fire the engines or to heat the cars during the severely cold Siberian winter. Discipline collapsed. Without leadership, disorder prevailed everywhere as people lost all hope of escaping the Reds' vengeance. Some committed suicide; the timid decided to take their chances and wait for the Reds; and the determined fled on foot, hoping to reach Irkutsk, a distance of 1,500 miles.

Irkutsk was the closest city that could offer the people protection, food, and rest. The majority, however, perished en route from hunger, severe cold, and typhus. Only those in good physical condition, possessing stamina as well as resourcefulness, reached Irkutsk. Had we been caught in a similar situation, I might have survived, but it is doubtful that Sonya could have withstood the hardships of such a journey on foot.

Later, in the city of Chita, when I met my former commander, Captain Astafiyev, he told the entire story of what happened to those caught in the tragic pandemonium. "I was riding in a train with the Tenth Aviation Squadron. We managed to pass Novonikolayevsk, proceeded a few miles east, and then stopped. The Reds were approaching Novonikolayevsk and east of us was an endless line of trains, hopelessly stalled and frozen to the tracks. Unable to move our train, we had no choice but to destroy our aircraft and equipment in order to leave nothing behind for the Reds to confiscate. We burned the works—everything! One of our men became so distressed and despondent he shot himself to death.

"A small group of us struck out on foot and headed for Irkutsk. Very shortly, we came upon an abandoned automobile, got it started, and drove it about ten miles before it broke down, probably because we had over-loaded it. We moved on, walking, but the arduous trek made its impact on our small group. Gradually our group disintegrated until Lieutenant Dudka and I were the only ones left. We tramped on through the snow and bitter cold, passed trains frozen to the tracks, many of them empty, some containing people resigned to their fate, others containing frozen bodies. Those still alive were dying of hunger, typhus, and other complications.

"Occasionally, we were fortunate to find someone with a horse-drawn vehicle willing to take us a few miles. Some peasants took pity on us, fed and housed us, and helped us on our way. Just before we reached Irkutsk, I noticed that Dudka showed signs of becoming ill. When we arrived in the city, I took him to the hospital directly. Since that time I have not heard from him, so I must assume he died there of typhus."

Captain Astafiyev was one of the few who survived to tell about his impossible trek to Irkutsk, and it was obvious to me that his ordeal had taken its toll. When a man endures prolonged hardships, witnesses extreme privation, suffering, and the death of countless war victims, his memories will haunt him the remainder of his life. Astafiyev's tale of woe, left me with the impression that the entire Siberian White Army was finished, but such was not the case. Some units of it had managed to retreat intact, and then gradually consolidated into a small army under the leadership of General Vladimir O. Kappel, thus providing some cover for the remnants of the White Army and civilians heading east.

All of the other White armies had suffered irreparable misfortunes. Their successes during the summer of 1919 had been short-lived. White units operating from Estonia were stopped in their march on Petrograd, dealt a decisive defeat on the outskirts of the city, and forced to fall back to Estonia. The White Army operating in the north at Archangel had been liquidated. And most important of all, the White armies operating in the southern part of European Russia had advanced within two or three hundred miles of Moscow where they had been stopped by the Reds and forced to retreat to the Black Sea and the Caucasus Mountains. Because of these massive defeats and reversals, the Allies refused to support the White armies, thus delivering the final, mortal blow to the movement.

Many acts of heroism took place as remnants of the White forces retreated from Omsk across vast distances between the widely separated settlements. Tales of their exploits became legendary, such as the one about the steel mill workers from the northern Uralsk. In the early part of the eighteenth century, Peter the Great established two large steel mills in the northern Uralsk Mountains for the purpose of manufacturing guns and ammunition. Over the years the mills grew into very large enterprises, and by the twentieth century, tens of thousands of workmen were employed there.

The mill workers refused to accept Bolshevik rule during the 1917 Revolution and staged a counterrevolution. They drove the Bolsheviks from their locality and defended themselves for a long time from constant attacks by the Reds until they were unable to withstand the repeated sieges. On November 14, 1918, they were forced to abandon the mills, and it has been estimated that about 30,000 men, women, and children left the area and headed south. All able-bodied men were organized into a well-disciplined army that continuously fought off attacks by the Reds as they retreated south to the Trans-Siberian Railroad. Here they eventually joined our Siberian Army and continued their fight against the Bolsheviks.

In 1919, during the general retreat of the Siberian Army, the Uralsk mill workers, who had become very experienced fighters, became the mainstay of the army, fighting until their remnants reached Irkutsk near Lake Baikal. On February 14, 1920, they crossed the frozen lake and proceeded on to Chita, a distance of 600 miles, fighting off the Reds all the way. Thousands perished in battle, from typhus, and the extreme cold. Only a few thousand of the hardiest reached Chita.

General Kappel, their talented commander, died tragically from gangrene just before reaching Irkutsk. During his long trek across the ice, his legs had become frozen. The White Army had intended to take Irkutsk, then governed by the so-called socialists and supported by the former Czech war prisoners en route to their own country. The plan had been to rescue Admiral Kolchak, the Supreme Commander of the White armies, until it was discovered that he had been shot by a Bolshevik firing squad a few days before. This act confirmed the suspicion that the "socialist" government in Irkutsk was in fact a front for the Bolsheviks.

Diary Excerpt, 1919:

Trek across Siberia with the White Army

The following excerpts from Alexander Riaboff's diary cover the period from June to October 1919. During this time Riaboff was a member of the Fifteenth Air Squadron, and he records in his diary the constant movement, mostly in retreat, along the Trans-Siberian Railway toward the city of Omsk. This section of the diary portrays vividly the rigors of the campaign, along with Riaboff's growing disaffection toward his commanders.

■　　■　　■　　■

June 19, 1919

Sometimes our forces stand in place for a while, even move forward on occasion against the enemy, but the Western Siberian White Army is in retreat. Our withdrawal is orderly, at about five-ten versts* each day.

Life in my squadron is fast-moving, but acceptable. There is little boredom. At times there is good cheer, especially when the general staff of the army issues orders for us to fly—often with little warning. Sometimes we are denied even a half-hour of peaceful rest.

Recently, Lieutenant Shamaniyel defected from the Reds to our side in his Nieuport 23 fighter. Captain Astafiyev at first promised this Nieuport to me, but once he saw the aircraft, he decided to fly it himself.

Shamaniyel tells me that he saw my name on a list of defectors that the Bolsheviks had compiled. He tells me that the Bolsheviks, according to rumor, are arresting and executing the families of defectors. O Lord, please save my family!

Shamaniyel also reports that the Reds are short of gasoline, so much so that they have concocted a fuel mixture of their own, using pure alcohol.** The Reds now are short of aircraft. They keep their pilots well behind the lines, fearing defections. Only Communists, or people they can trust, are permitted to fly.

---

*A pre-revolutionary Russian unit for the measurement of distance, a verst equalled 0.663 miles or 1.067 kilometers.

**Called the Kazan mixture, this make-do fuel consisted of alcohol, ether, toluene, and other additives. It failed to supply full power, but did serve as a workable expedient, although it corroded metal and hoses.

June 23, 1919

Yesterday, I flew a reconnaissance mission again, covering a route in the Asha-Balashevskaya region, near the River Ufa. . . . This reconnaissance flight proved rather disappointing; flying at dawn, I was over mostly forested areas for 2½ hours, too early to spot anything.

The aircraft assigned to me has been tested twice, and the engine, which first gave me concern, is now working well.

Our air unit has been assigned to the Asha-Balashevskaya region. We operate from a small and poorly equipped air strip. Despite these primitive conditions, we continue to fly without a problem.

The Reds have crossed the Belaya River and have occupied the city of Ufa. They recently crossed the Ufa River as well, but soon ran out of steam. Now they appear stalled, as they draw back troops for reassignment outside Petrograd to defend against General Yudenich. He already has occupied Krasnaya Gorka, which is about twenty versts from Petrograd. Meanwhile, General Denikin is on the move in the south, having crossed the Donets River Basin and taken a position near Tsaritsyn, which is still held by the Reds.

Our own Army of Western Siberia has been on the offensive first to occupy Glazov and now on the offensive to take Vyatka. We have had excellent success against the Reds, especially in the Mendelinskiy direction. The Cossacks in the Ural region have been moving forward against the Reds and we are now within sixty versts, approaching Saratov. . . .

July 1, 1919 *(at the Kropochevo Station)*

For four days we have been at Kropochevo, about fifty versts east of Asha. On June 26, I made a successful reconnaissance flight with solid intelligence material gathered on the enemy around the Asha-Balashevskaya and Ufa River areas. . . . This flight was very difficult because I had to fly much of the route without familiar points of reference such as roads, rivers, or other landmarks. For proper orientation, I relied on the sun, knowing only that on my left side was the Ufa River and on the right side the mountains. . . . With luck, I found my way.

My return to base was direct, but not without danger. My fuel supply was low, and, for the last fifteen versts before landing, I had to fly over rough, mountainous terrain. If at any point the engine had stopped, I would have dropped into the thick mountain forests below. I found myself praying fervently to God for the meager fuel supply to hold out. I believe God heard my prayer. . . .

American airplanes are now reaching our army. The Fifteenth Air Unit, I understand, received three of these aircraft. Our training school also received another three airplanes. What kind they are, I do not know. . . .

The White armies, even our Western Siberian Army, are now in retreat, although the Ural Cossacks continue to advance against the Reds.

There have been numerous arrests for black marketeering, including the arrest of Colonel Boin-Radziyevich.

August 22, 1919 *(at the city of Omsk)*
Our air unit has been reassigned from Kropochevo to Mursalimkino. My flight to the new air base did not go well. On takeoff, my engine stopped suddenly and I had to land downwind, damaging the undercarriage of my aircraft. After making the necessary repairs, I then flew to Mursalimkino. No sooner had I arrived at this new base than I was ordered to fly north of the railroad line, to look for the enemy. . . .

Our position here is one of extreme danger. The Reds have continued to advance in this region and have threatened to encircle us. Rather than be cut off, we disassembled our aircraft and moved them eastward along the railway to Miyass, to a point beyond the reach of the Bolsheviks. Once there, we reassembled our airplanes and established our new air base. I received a two-week furlough, returning to my air unit in mid-July at the village of Medvedev. While at Medvedev, I flew several reconnaissance missions north of the railway. I remained four days with the Second Air Unit, where I picked up an aircraft for my own squadron. I reached Omsk on July 19.

While in Omsk, I have seen our aviation command up close. My impression of our leaders has not been pleasant—heavy drinking, few progressive ideas, no real interest in aviation or in finding practical solutions to our problems, all kinds of self-serving excuses for failures.

Our retreat continues—we are falling back, abandoning the cities of Miyass, Chelyabinsk, Kurgan, Glazov, Perm, Yekaterinburg, and so on. We set up a defensive line at the River Tobol, only to see Reds cross it at several points.

All White air units are now in Omsk. . . . including my own. Work continues in the repair shops on our aircraft, especially to recondition our aero engines for the upcoming struggle.

Here in Omsk I have met my old friend Devoyod, who escaped to the Whites with me last year. He tells me that all of us who escaped— Yefrimov, Devoyod, and myself—are now on a wanted list with a reward of 2,000 rubles offered for our capture. Strange as it seems, the latest copy we obtained of *Vestnik vozdushnogo flota* [organ of the Red Air Fleet] still lists us as active pilots on their planning board list!

September 8, 1919
Along with Ensigns Volkovoinov and Katalnikov and several motormen, I left Omsk on August 25.

My squadron has received three Sopwith aircraft from the French,

The Retreat of Kolchak

[Excerpt from George Stewart, *The White Armies of Russia,* Macmillan Co., New York, 1933. Reprinted by permission of Robin Higham, literary executor.]

B y late September, Kolchak's army had lost all semblance of a fighting force and had become a mass of refugees fleeing toward Lake Baikal, overcrowding the towns, paralyzing such transportation in Western Siberia as the Czechs did not control, and filling the towns with dispirited troops. "The army," writes Miliukov, "was running eastward, abandoning supplies and equipment on the way, fearing nothing so much as to be overtaken by the Reds, and putting tens of versts between themselves and their pursuers."

The forces were suffering heavily from frost, lack of food, disease, disorganization, and the attacks of the oncoming regiments of the Red Army. Omsk, where Kolchak's retreating soldiers had crowded together in a stampede to escape the Red Army, was a lonely city in the midst of a vast steppe surrounded by Tartar villages. As Kolchak's fall became more and more imminent, the normal population of one hundred and twenty thousand was swollen to well over half a million. Every building was crowded, full of soldiers and refugees, dirty, lousy, infected with spotted typhus and typhoid fever.

When the main body of the troops arrived at Omsk, they found unspeakable conditions. Refugees overflowed the streets, the railroad station, and public buildings. The roads were hub-deep in mud. Soldiers and their families begged from house to house for bread. Officers' wives turned into prostitutes to stave off hunger. Thousands who had money spent it in drunken debauches in the cafés. Mothers and their babies froze to death upon the sidewalks. Children were separated from their parents and orphans died by the score in the vain search for food and warmth. Many of the stores were robbed and others closed through fear. Military bands attempted a sorry semblance of gaiety in the public houses but to no avail. Omsk was inundated in a sea of misery.

The condition of the wounded was beyond description. Suffering men often lay two in a bed and in some hospitals and public buildings they were placed on the floor. Bandages were improvised out of sheeting, tablecloths, and women's underclothing. Antiseptics and opiates were almost non-existent. Only in the larger bars and cafés did there seem to be money and food and drink. Winter had set in and hundreds were dying of exposure.

Day and night, roads leading to the east were jammed with horses, camels, and donkeys dragging guns, sleigh, and carriages. Men, women, and children stumbled over the snowy waste leaving crimson stains on the road as a testimony of their suffering. No serious officer could hope to reform the sullen and dispirited troops.

equipped with 130-horsepower Clerget engines. On August 30, I got to fly one of these aircraft with a passenger, Lieutenant Moshkov, for thirty minutes. On August 31, I flew a frustrating reconnaissance flight close to the enemy positions near Makushino. It was not very successful—low clouds and poor maps prevented me from establishing contact with the Reds.

Another reconnaissance mission flown on September 1. This flight proved to be very successful. I discovered one detachment of Red troops in the area of Bronov. Successive reconnaissance missions over the next two days enabled me to establish contact with other Red detachments retreating westward. I've logged over thirteen hours during the past four days. . . .

At present the Reds are falling back, or, to be more precise, we have them on the run! The deepest point of enemy penetration came when they occupied a point east of the Petukhovo station. That was on September 1. Yesterday, we received word that our troops had pushed the Reds back a total of ninety versts, as far as a point west of the Lebyazhiye station. Our troops continue to advance in the north, but we continue to fall back in the region south of the railway.

At this time my squadron has three Sopwith aircraft, one Parasol, and three Nieuport 23s. The squadron roster shows seven pilots for seven aircraft: Captain Muromtsev, Lieutenant Dudka, and Ensigns Volkovoinov, Yelsner, Katalnikov, Plishke, and myself, Ensign Riaboff.

September 14, 1919 *(at the Petukhovo station)*

Five days have passed since we arrived in Petukhovo. Volkovoinov and I ferried the Sopwiths here. We were joined later by Captain Muromtsev and the rest of our squadron.

I've made one reconnaissance flight since my arrival, to the Lake Kurtak area. This mission proved to be difficult, with continual engine problems plaguing me. I had the engine replaced, but on a reconnaissance flight near Makushino I faced problems with the new engine, compelling me to make a forced landing. They are now working to repair my old engine. . . .

Our army continues on the offensive. We have taken several thousand prisoners, along with 100 machine guns and other military equipment. The Reds are now planning to counterattack. The future is uncertain. . . .

September 25, 1919 *(at the Petukhovo station)*

The Reds have counterattacked, but we have held our lines. They are falling back again.

My aircraft is now repaired and the engine is running well. I have grown dissatisfied with my squadron, and I have requested a transfer to another air unit. My reasons are many, but Captain Muromtsev's arrogance, and the heavy drinking in the squadron, stand out as my chief complaints. . . .

Muromtsev, at first, refused to approve my transfer, but he finally approved it, despite the shortage of pilots.

Drinking among my squadron mates has reached a new level, which has caused me great concern. . . . Lately, I have become very nervous and tense. My doctor tells me I'm a nervous wreck. I look forward to my next furlough!

Good news from the west. General Denikin has occupied Chernigov and Kursk.

September 29, 1919 *(at Makushino)*
Today we flew over Makushino. I was going on a reconnaissance flight today, but poor weather forced me to abandon the mission. I did fly escort for Lieutenant Yankovskiy on a short flight.

October 23, 1919 *(at the city of Omsk)*
After leaving Makushino, we moved to Lebyazhiye, and then to Vargashi, where we remained for two weeks.

While Vargashi, we made contact with a Red fighter, a SPAD, which attacked Volkovoinov. At that time I was flying escort and I attempted to give chase, but the enemy escaped.

On a night mission Captain Muromtsev destroyed a Red balloon over the railroad station at Zaryanka. Voshchilio was wounded in this attack.

Finally, I left my squadron on October 7, making the trip to Petro-pavlovsk to join the Fifteenth Air Unit. When the official orders came, they indicated that I should report to Atbasar in Turkestan! Some of my so-called "friends" in the squadron took delight in my posting to this remote place. But their joy was premature. When I departed I took the precaution of going to Omsk first, to confirm my assignment. This decision proved to be timely, because at Omsk I found my new commander, General Kompa-niyetsev, who told me to stay with him and to wait for the 15th Air Squadron, which was on its way from Atbasar to Omsk. This news gave me considerable pleasure!

Our troops have been thrown back from the Tobol River, for about thirty versts, in a Red counterattack. Still good news from the west, where Denikin and Yudenich continue to enjoy success against the enemy. Our troops there have taken 9,000 prisoners.

Prices have risen rapidly in Siberia, at least 50 percent in the past two weeks. Compared to a year ago, prices are now ten times higher. Butter costs two rubles a pound, as opposed to 1.8 rubles earlier, milk is five rubles, a one-fourth liter of wine about two-three rubles. Paper money is nearly valueless.

Captain Duklau, who commanded the Fifteenth Air Squadron, and an observer, Captain Borovskoi, have died.

8. Siberian Struggle

When the Kolchak command at Omsk fell to the advancing Red Army in November 1919, Alexander Riaboff joined the White forces in another grim retreat eastward. There was little the Whites could do at this juncture of the civil war to halt the inexorable advance of the enemy.

As before, there were still tactical victories for the Whites in this second major retreat, but they were less frequent. The political disintegration of the White cause became more evident to Riaboff. There was continued social upheaval behind the frontlines, with an alarming rise of defections to the Red side, and declining morale among the White troops.

Since 1918 the White offensives, one by one, had been turned back. By 1920 no leader had emerged in Siberia to replace Kolchak, and the Allied interventionists by this date showed less sympathy toward the Whites, except for the Japanese, who were anxious to use unscrupulous types like General Grigoriy Semyonov for their own purposes. The year 1920 brought untold hardship to Riaboff, who at the end of it went into exile. By 1922, the last of the White armies had withdrawn from Siberia, along with the Japanese troops, ending one of the most brutal military campaigns in history.

■　　　■　　　■　　　■

WHEN IRKUTSK FELL IN 1920, we faced another difficult retreat eastward along the Trans-Siberian Railway. The White Army, now consisting of about 5,000 men, proceeded toward Chita.

Captain Grishin and the personnel on our train agreed unanimously that the cause for which we had been fighting was irretrievably lost. Therefore, they chose to head straight to Shanghai and settle there. Following a lengthy discussion, my wife and I chose instead to remain in Chita, where I hoped to join the local aviation squadron and continue the struggle.

Near Chita, General Semyonov's army had one aviation squadron, which welcomed me as a pilot, although it lacked airplanes and equipment

Members of General Grigoriy Semyonov's squadron in Siberia. Semyonov used his air unit to help establish his control of portions of the Trans-Siberian Railroad.

and was under the command of a Captain Pleshkov, who was not a pilot. Nevertheless, the squadron had two noncommissioned pilots who informed me that they expected the arrival of French aircraft from Vladivostok in the very near future. The Semyonov army, including the aviation squadron, leaned toward the extreme right of the political spectrum. It was in favor of restoring the monarchy, the old laws and social order, and hated any form of liberalism. Semyonov's harsh regime did not allow any criticism to be voiced against it, and any infraction of this rule could lead to tragic consequences, as I was to learn sometime later.

The remaining Siberian army did not merge with Semyonov's group, but by mutual agreement, Semyonov took on the defense of the eastern section of the Trans-Siberian Railroad, while the Siberian army defended the western section from Irkutsk to Chita. Each army retained its separate command. Eventually, some old American biplanes arrived in Chita from Vladivostok.

Ensign Mozhevitinov, a pilot who had joined our squadron, tested one of these aircraft with unfortunate results. It was sub-zero winter weather and the airplane had a water-cooled engine. Hot water was poured into the radiator several times in order to warm the engine and the pipes connecting the radiator with the engine. Each time, the water was permitted to flow out until the motor became warm enough to turn over, then the radiator was filled and plugged.

Finally, the pilot took off, circled the field once, and appeared to be coming in for a landing. His approach to the ground was erratic and we knew something was wrong, either with the airplane or its pilot. He crashed-landed, completely demolishing the aircraft, but without injury to himself, except for a few minor scratches.

Mozhevitinov was a bit dazed when we helped him out of the wreckage, but he was perfectly able to explain what had happened. "A few moments after I took off, the water in the radiator froze and circulation to the engine stopped, causing water around the engine to vaporize. Steam settled on the windshield and on my face, then froze. My eyes were frozen shut and I had to land! I'm surprised that I'm still alive.!" I told him that he was no more surprised than we were.

While I was serving with the Semyonov squadron, Sonya and I lived close to Chita in a small village by a lake. It was only a short walk to our place from the flying field with its shops and hangar. On one extremely cold day, I had gone on an errand that necessitated my being outside in the sixty below zero Siberian weather somewhat longer than usual. It was a beautiful, clear, sunny day which caused the snow to be blindingly white. The air was still. Smoke from chimneys formed straight, high columns, dissolving into nothingness at the top. There was no one in sight to disturb my reverie or the prevailing silence as I beheld the sparkling elegance of a Siberian winter.

I walked on slowly, breathing carefully, as the bitingly cold air burned my nasal passages. I was covered with heavy clothing from head to toe, only my face completely wrapped in a woolen scarf, leaving only a narrow slit for my eyes. When I arrived home and removed my scarf, Sonya observed, "My goodness, the bridge of your nose is frozen!" And, as I continued to peel off my layers of clothing, I became aware of an unusual feeling in a very vital area of my lower torso.

A cursory examination revealed that material evidence of my masculinity seemingly had disappeared from sight. "What has happened to me?" I wailed, horror-stricken, wondering if I would ever be the same man again. A palpitation in this same area indicated that the recalcitrant organ had submerged itself into my body, and my efforts to elicit its reappearance failed completely, leaving me perplexed, to say least.

After a half hour of chagrin and apprehension, my normal self was restored without any effort on my part, proving that nature is far more ingenious than one might suspect. Nature had outsmarted someone who thought he knew everything there was to know about his own anatomy. Yes, indeed, I remember that winter day in 1920 only too well!

Soldiers from our squadron visited the local village quite often and became friendly with some of the people. The officers discouraged fraternization with the locals because they suspected there might be Bolsheviks lurking among them. On one occasion, late in the evening, a violent fight erupted between a group of our soldiers and some of the young men in the village, presumably over attentions being paid by members of our unit to the local girls. Our soldiers were the losers, and retreated to their barracks bearing all the visible signs of defeat. Word of the incident spread throughout the entire squadron the next day and our boys took a severe ribbing for having lost the battle with the local civilians.

When our commander heard about the episode, he took it very seriously and immediately appointed a committee to investigate the details of the affair. In reviewing the incident, the committeemen found that our soldiers were innocent of misbehavior and that the village boys had been the villains and should be punished. The commander sent a report to the staff in Chita concerning the altercation, along with the committee's findings, and requested instructions as to how to dispose of the matter. "Do whatever you find is necessary," the staff responded. Without further delay, a dozen or so armed soldiers were ordered to go to the village to locate and arrest the assailants.

By that time those involved in the fight had either left the area of were in hiding, because the soldiers were able to apprehend only seven young men, all of whom declared their innocence of wrongdoing. Nevertheless, they were brought to trial before a military court made up of squadron officers. I happened to be at headquarters at the time a group of officers and the commander were emerging from a meeting room. I could see by their dour facial expressions that they had been discussing something of a very serious nature. The commander turned to me and said, "We have decided to try the arrested villagers tomorrow. Do you want to serve on the military court?"

By the general attitude of the group I was sure the fate of the defendants had been determined already and my efforts to try to save them would be in vain. I knew that this tribunal was out to get Communists and to teach the villagers a lesson. If I were to express a pacifist's opinion regarding the affair, I could incriminate myself and then they would accuse me of being a Bolshevik sympathizer, which would seal my fate as well.

"No sir. I would rather not serve in this case," I replied.

He continued, "So, you would rather not serve in this case, Ensign Riaboff? I don't like your attitude. You had better shape up—change your way of thinking and your political beliefs, or else!"

I knew that he meant his reprimand to have been a stern warning and I shuddered at the possible implications of his threat. During the trial the peasants vigorously denied that they had participated in the alleged altercation and insisted they had been absent from the village that particular evening. As I had foreseen, the seven were found guilty of assaulting the soldiers. But, most important of all, they were found to be Bolsheviks and were given the death penalty without appeal. Because it is impossible to dig graves during the Siberian winter, the executioners decided to take the seven prisoners to the nearby lake, have them cut a hole in the ice, shoot them, then shove their bodies through the hole under the ice.

When the officers contemplated the actual operation of their plan, they became so unnerved by the grimness of it, they had to fortify themselves with vodka. In fact, they fortified themselves so well, they were unable to shoot accurately. As a result, they had to shoot their victims repeatedly in their attempts to kill them, and some, I was told, cried out as they were shoved alive into the hole in the ice. The entire episode was an act of injustice at its worst—inexcusably brutal and sadistic.

When the people heard what had happened to their boys, they were horrified, and from then on no soldier dared to venture anywhere near the village. I, too, was horrified, ashamed, and beside myself with remorse for having refused to take part in the court hearing. For days I was plagued with guilt feelings, yet I knew in my heart that my efforts to dissuade my fellow officers from executing the prisoners would have been useless. Their fate had been sealed from the time of their arrests because it had been

Grigoriy Mikhailovich Semyonov 1890–1945

Grigoriy Semyonov led a group of irregular troops against the Bolsheviks between 1917 and 1921. Known for his extreme brutality, Semyonov controlled a large portion of the Trans-Siberian Railway east of Irkutsk during the time of Alexander Riaboff's trek across Siberia. At his peak, Semyonov had over 60,000 men in his army. Much of his support came from subsidies from the Japanese interventionists who saw in Semyonov a convenient tool to exert control in Siberia. While under the nominal authority of Admiral A. Kolchak in 1919, Semyonov effectively abandoned the military struggle of the Whites to pursue his activity as a bandit and freebooter.

Semyonov had graduated from the Orenburg Military School in 1911 and served briefly in the tsarist army against Germany in World War I. During the revolutionary upheavals of 1917, Semyonov took up the cause of counterrevolution in Siberia. His troops terrorized the local population along the eastern part of the Trans-Siberian Railway, exacting tribute, looting and burning villages, and executing numerous people, sometimes entire villages, under the pretense of suppressing Bolshevik agents and sympathizers. For many honorable officers and soldiers fighting the Bolsheviks in the civil war, service in Semyonov's army presented numerous ethical problems. Semyonov's behavior, as they complained, did much to undermine the political and military position of the Whites.

For a time, Semyonov cruised the Trans-Siberian Railway in his two armored trains, the "Merciless" and the "Destroyer," creating considerable havoc and alienating the populace against the White cause. For these actions he won the intense dislike of General William Graves, the commander of American interventionist troops in Siberia, who considered Semyonov and his allies as hirelings of the Japanese and murderers.

General Semyonov escaped to Manchuria in 1921. He attempted to settle in the United States, but was refused because of his Siberian exploits. He then spent his life in exile, living at various times in Korea, Japan, and Manchuria. When Soviet troops invaded Manchuria at the close of World War II, Semyonov lost his Japanese protection. He was captured in September 1945, taken to the Soviet Union for trial, and later executed.

decided by the command that they were Communists and "all Communists must die"—in what other way could the villagers be taught a lesson!

I doubt whether the simple folk in that remote Siberian village had ever head of Karl Marx—or Lenin, for that matter. After all, they lived thousands of miles from the center of political life. I have often contemplated just how close I came to being tried, shot, and having my body shoved through a hole in the ice.

With the arrival of two new pilots, Captain Pavlov and Ensign Agapov, our personnel increased favorably and soon afterward we received the two promised French aircraft. They were sturdy, reliable biplanes with powerful radial engines of 230 horsepower! In addition to the pilot's cockpit, there was a large two-seater one in the rear. I was ordered to test the aircraft, and then make several routine flights, after which I continued to fly it regularly.

Once, on a reconnaissance mission over a desolate mountain range, I suddenly detected the odor of burning rubber. I knew it must be coming from the engine and I expected the motor to stop running at any moment. Looking down, I saw there was no place to make an emergency landing and detected no signs of civilization for miles in any direction. As I agonized over the hopelessness of the situation, the motor continued to run until I completed my flight and I landed normally. Upon investigating the cause of the odor, my mechanic found an electric wire that had sagged downward, touching the cylinder and burning its insulating rubber. Luckily, the motor had double ignition wiring, which had saved the airplane and its pilot.

Our squadron moved to Daurya, a railroad town located 100 miles west of Manchuria. At that time, Daurya was the headquarters of the notorious Baron Ungern von Sternberg. A legendary figure of the Far East, Ungern, who was of German descent, had been commissioned a midshipman in the Imperial Russian Navy, but he detested going to sea and proved it by joining a Cossack regiment stationed in Siberia at the Mongolian border, far removed from the sea. He became an expert horseman. To add to the complexity of the man, he also became known for his avid interest in Buddhism. During the civil war, he organized a special Cossack division to fight the Communists and became warlord of the region from his headquarters in Daurya. His reputation as a brutal disciplinarian had spread throughout Siberia. Ungern kept his division in a constant state of fear by threat of punishment for the slightest violation of discipline.

An accused offender was sent to the roof of the central barracks and was ordered to remain there until further notice. In the dreadful cold of winter or in the scorching heat of summer, this manner of punishment could be fatal. Occasionally, the cook would toss some food up to the victim to keep him alive, otherwise he would have died of starvation. Another way in which Ungern punished his men was to strike them on the head with a light club that he carried with him at all times.

He devised a clever way to support his division by levying a duty on goods and people passing through Daurya. Anyone refusing to pay was severely punished and, of course, all persons suspected of being Communists were shot on the spot.

One day, I was preparing an airplane for a reconnaissance flight when a crew member announced in a trembling voice, "Oh my God, Ensign, Baron Ungern is coming toward us!" I glanced up to see a thin, poorly dressed, lone horseman riding in our direction. I waited for him to draw near us, brought my men to attention, and walked stiffly, in the Russian military manner, right up to the rider and saluted him. "Ensign Riaboff reports; he and his crew are preparing this airplane for a reconnaissance flight, sir."

He returned my salute and inquired, "What's the make of the airplane?"

"A French type with a 230 horsepower Salmson engine sir," I responded.

"Where are you headed?"

"North, sir, in the direction of Nerchinsk, then southwest to the Trans-Siberian Railroad and return."

"How long will the trip take you?" he asked. "About three and one half hours, sir," I responded. "Good luck, Ensign." He smiled at us, turned his horse about, and rode away. "Thank you sir," I called after him, with a genuine sense of relief. My crewmen looked at me with expressions of gratitude as if I had saved their lives. Such was the reputation of the formidable Baron Ungern.*

On May 30, 1920, Colonel Kachurin, my new commander, Captain Slusarenko, Ensign Agapov and I were sent to Nerchinsk. It was a small town 150 miles from the Chinese border and 200 miles north of the Trans-Siberian Railroad, with a connecting railway from the main line. We made numerous reconnaissance flights to the north and west of Nerchinsk, where we spotted many Red Army units.

As a result, our air staff wished to show its gratitude for our valuable discoveries by honoring us with a banquet. A very wealthy merchant who lived in the town offered his residence for the affair. It was a beautiful home with the finest of furnishings, none of which I recall in detail except

---

*Baron Ungern von Sternberg's harsh leadership prompted some of his own troops to mutiny in June 1921. They turned him over to the Bolsheviks, who took him to Red-controlled Irkutsk in a cage. He was executed that same year.

for one unusual item. Hanging on the living room wall was a large mirror, approximately ten feet long and five feet wide, mounted in an elaborate French baroque gold frame. I had never seen its equal in the museums or palaces in Moscow or Petrograd. I stood before it, struck with the incongruity of finding such a remarkable possession in a town without a single paved street or sidewalk. My curiosity to learn of the mirror's origin and how it had been transported to its destination prompted me to find our host and ask him. I thanked him for his hospitality and said, "Please tell me the story of how you acquired such a remarkable mirror. I have never seen a more perfect one in all my travels."

"I'll be happy to tell you about it." he replied cheerfully. "I bought it at the 1905 World Exposition in Paris. It was crated and taken to the nearest railroad station, and then it traveled by rail across western Europe to European Russia; from Moscow to the Trans-Siberian Railroad. There was no branch line to Nerchinsk at that time so from a certain stopping point on the railroad I had to hire several dozen local peasants to hand carry it nonstop the fifteen miles to my home."

"Thank you, sir, for telling me about your and Nerchinsk's most unusual work of art," I said. "Someday I'll repeat your story to my children." When the Reds captured the town, they took possession of the merchant's grand home; I have wondered at times if they really appreciated that mirror.

While I was flying in Nerchinsk, Sonya and I lived in an especially equipped boxcar stationed on a spur track outside of the town. One evening, word reached us that a large Red Army unit was moving in the direction of our location. We were told to be ready to leave the train and prepare ourselves to retreat south through a rugged mountain chain to the Trans-Siberian Railroad.

"Take as few things with you as possible," our informant advised us. I said to Sonya, "Pick up a few pieces of essential clothing and forget the rest." And I left her to go to my fellow officers and discuss the situation with them.

When I returned, I found a very plump, blond, blue-eyed tea cozy waiting for me that hardly resembled my wife. Lying on the bed was a very large bundle containing her idea of essentials for a foot journey through the mountains. "Well, my darling, I see that you have packed everything except those two chairs over there, which we are sure to need when we sit down to rest in the mountains!"

"By the way. I'll give you the pleasure of carrying the bundle." She looked at me quizzically, wondering if I were serious, and when she attempted to pick it up I put my arms around her sizable waist and said, "Don't worry, we aren't going anywhere tonight. It was a false alarm." We both laughed with relief as she pulled off several layers of "essential" clothing and reduced her weight about thirty pounds in a few minutes.

On June 11, 1920, I was ordered to fly north over the mountains to deliver an important order to General Molchanov, the White regimental commander operating from a small village nestled between large, surrounding mountains. My navigator and I had difficulty locating the village, but eventually we found it, only to be confronted with the problem of where to land.

I finally decided to bring the airplane down on a level clearing right in the village. On approaching the landing site, I discovered that the clearing I had seen was a diminutive plateau, level on top and covered with tall grass. Having no alternative. I took the risk of landing there. My wheels touched down on the edge of the hilltop, ran through the tall grass, and the aircraft headed for the down-grade on the opposite side.

"Oh my God, we're going to turn over!" I shouted to my navigator. "Jump out and try to stop the airplane!" At that moment the aircraft came to a halt because the tall grass substituted for the brakes we didn't have. We climbed out of the airplane, looked around, and saw a man on a horse approaching us, and a number of men in uniform running toward us. The rider reached us first, dismounted, and stood before us. He wore a military tunic, was little more than a teenager, and when we looked at his shoulder straps, we became momentarily speechless. He was wearing a general's insignia and was entitled to the respect due his high rank.

My thoughts skipped from one line of reasoning to another. Was this youngster a real general? Misrepresentation of one's rank was severely punished, and it was unheard of to make a joke of such a serious matter, so he must be entitled to his insignia, I argued mentally. Then I remembered a story of an extremely brave high school boy who had joined the White Army at its inception and had been rapidly promoted to a general's rank, being referred to as "the student-general." Without further hesitation I stood at attention, identified myself, and stated the purpose of our coming. Before reporting to General Molchanov, I asked the young general, "If you please, sir, we will need some help to pull our airplane back to the edge of the hill. I will need all of the runway available for takeoff.

He ordered several of the soldiers to comply with my request and to remain on guard with the aircraft while my navigator and I were led by a small company of privates to General Molchanov's residence, an ordinary peasant's cottage, the only type of housing found in this mountain valley. Word of our arrival had already reached the general, and he appeared to be very pleased to see us.

"I am Ensign Riaboff, sir, sent by the general staff. Here is the order designated for you." I presented the package, which he opened immediately, quickly reading its contents. When he was finished, he looked up at me and said, "This is exactly what I wanted to know. Thank you, Ensign. I'll give you my answer in the morning." After the general's reply was delivered to us the following day, we proceeded to our improvised airfield, trampled down the tall grass, started the airplane, and took off.

As soon as we returned, we delivered General Molchanov's package to General Lokhvitsky personally. General Lokhvitsky heard my report of our flight with great interest, interjecting detailed questions about flying, and then he said, "Ensign Riaboff, brief me on your education and your military service, please." After I outlined this information, he said, firmly, "You should have been promoted to the rank of first lieutenant." He took a deep breath and continued, "I hereby promote you to the rank of captain for "extremely important service to our country!""

General Lokhvitsky then questioned my navigator, put him through the same procedure, and promoted *him* to the rank of captain. Promotion orders were drawn up immediately, were signed by all parties, and two additional stars were drawn in indelible pencil on our shoulder straps. Although the civil war was coming to a close, and our promotions had no practical value, we were happy to have had our services recognized and rewarded.

We returned to squadron headquarters and officially reported our promotions to the commander, Colonel Kachurin, who did not react favorably to the news. He was surprised to learn that we had been promoted in rank as the result of our flight to General Molchanov. "Congratulations gentlemen," he said reluctantly, with no joy in his voice. "Well, I guess you must have earned your promotions." All of the other officers congratulated us and we had a brief celebration to commemorate the event.

By July 1920, our squadron had moved to Borzia, a station about 100 miles from the Manchurian border. By now the military situation for our fragmented White armies had become desperate. Train followed train heading east with retreating Russians. By such indications, I predicted that the civil war could not last more than a month or two, at least in our locality. We knew we would be forced to leave Siberia for Manchuria in the near future.

On August 10, I was ordered to fly to Chita to procure foreign passports for everyone in the squadron so that we would be prepared to travel abroad. In two days I returned with the passports and distributed them. To add to the continuing hopelessness of the situation, a group of four pilots and a navigator defected to the Reds in four of our aircraft. Such an occurrence was unprecedented, and when some of the military units defected, I concluded that further efforts to prolong the war were useless.

By this time the Russian ruble was worthless and could not purchase

Alexander Riaboff, seated on the running board at the right, joins members of his squadron for an informal picture during the 1919 Siberian campaign.

anything. Fortunately, our salaries for July and August had been paid in gold and silver coins of Imperial Russia, and I now had in my possession seventy-two rubles, all in valuable coins. I was well aware of the fact we would have to have negotiable money when we entered another country. Because of our rapidly depleting forces, our squadron even accepted several former criminals who volunteered to serve with us.

All went well until August 21, when, at midnight, five of these same volunteers took flight, taking with them our horses and rifles. As soon as I heard the news, I offered to aid in the search for the culprits. My airplane was under repair so I requested the use of our absent commander's aircraft. I was given permission to use it by his substitute, who was in charge. Two armed officers and I flew in a northerly direction to the mountains, where the Reds were rumored to be stationed. I spotted the horses about twenty miles away from Borzia and I knew that their riders must be close by. Immediately, without caution, I descended to get a closer look at the area to determine whether or not the men were hiding out there. I came within fifty feet of the ground, pointing my machine toward the horses and, sure enough, three men were lying behind some boulders with their rifles aimed directly at the aircraft. Just as I approached them, I could see the one man had his gun pointed directly at me from a distance of less than 100 feet. I pulled the gas lever to its maximum; the airplane lurched forward and at the same moment the man fired and I heard the bullet hit. Subsequent shots missed the aircraft, but one of my passengers shouted, "The fuel tank has been hit!"

The fuel tank, located between my cockpit and theirs was emptying its contents rapidly, but the airplane was equipped with a spare tank and we had no difficulty on the return trip to our home airfield. After I had landed the airplane and made a quick report, the commander ordered a detachment to pursue the escapees. But this mission was in vain, for they were nowhere to be found. One of the pilots in the squadron came to me a few days later, after the commander had returned and discovered that his airplane's tank had been shot through, and reported: "Colonel Kachurin is mad as hell at you for letting his aircraft get shot up while he was away!"

"Well, at least the three of us came back alive," I said. "If I hadn't jerked the airplane forward in time, I would have been shot right in the chest. One second later and a bullet would have hit one or both passengers. After all, the gas tank was repaired in two days' time and other than that, the aircraft was not damaged in the least."

I was angry at the colonel's concern for his aircraft and his indifference to our having just barely escaped with our lives in the line of duty. "You have my permission to repeat what I said," I told my colleague as he walked away. Once again, luck had been with me, and perhaps I was guided by the same "angel" that had kept me from destroying my airplane and myself on many occasions; from boarding the wrong train; from going to Uralsk; and best of all, for arranging my introduction to Sonya, my helpmate in every respect for more than twenty-five years.

One day while I was stationed in Borzia, my former squadron-mate, Lieutenant Voshchilio, appeared out of the blue. His constant companion and my former commander, Captain Muromtsev, was not with him, so the first words out of my mouth were, "Where's Muromtsev?" Voshchilio's expression changed immediately to one of pain. I should have weighed my question more sensitively, I thought, and not have been so quick to inquire. Considering the fact that I had not seen either man since the summer of 1919, Muromtsev could have lost his life.

Voshchilio struggled with his answer to my question, and after a long pause said, "He has gone to a far better place, and I've often wished I had gone with him." I then said, "I'm sorry, my friend, I shouldn't have asked about him the way I did, but you two were always inseparable and I expected to see him following closely behind you."

Several men in the squadron had gathered around us and had listened to our conversation. "The last time I heard about you and Muromtsev was in Omsk. I was told that you had not returned from a reconnaissance flight over Red territory and no one seemed to know your fate. Please tell me what happened," I asked.

"I came here especially to have a reunion with old friends and to share my past experiences with them. What happened to us is a long, sad story," he said.

Those of us who had formerly known pilot Muromtsev and his navigator, Voshchilio, listened eagerly as he proceeded to relate the following drama:

While flying over Red territory, we were forced to land when our motor stopped suddenly. There was a forest nearby, so we jumped from the airplane and tried to run into the forest and hide, but some Red soldiers had watched us land and were waiting. They chased us a few yards, caught us, and gave us a sound beating. They would have beaten us to death if their aviation personnel hadn't arrived on the scene when they did.

The pilots realized who we were and ordered the soldiers to stop. When one of the Communists, probably a commissar, shouted, "I demand that we execute these White dogs immediately," the pilots rushed to us, began to shield us from further abuse, and one of them said, "You can't shoot these men now, they are half dead already. We'll take them to the hospital."

Before the commissar could act, the pilots picked us up and carried us to an automobile and took us to the hospital, where we received excellent care and were ready to be discharged after two weeks. In the meantime, our pilot friends were plotting a way to save our lives. They knew the commissar was a fanatical Bolshevik with a long memory and would not forget his threat to execute us. The doctor, who treated us, very cautiously hinted to the pilots that our dismissal was imminent and unless a way was found for us to become seriously ill, particularly with a contagious disease, such as typhus fever, he could not detain us any longer as patients in the hospital. If we contracted typhus, we would have to be evacuated way to the rear of the Red Army forces and the vindictive commissar. "You have a 50-50 chance to survive typhus, but if you are dismissed from the hospital, your chances to live are nil," we were told. With such an alternative, we decided to take our chances with typhus.

We were directed surreptitiously to a certain room and were told to sleep in the beds where two men had died of typhus two days before. Within a short time we became very ill, were removed from that hospital, and were sent under assumed names to a hospital in a city hundreds of miles west of where we had been. From that hospital, we were sent still farther west and again with new names.

Eventually, we recovered completely and were released as invalids, unfit for military service. It's true, we were painfully thin, pale, and far from the rugged specimens we once were, but we felt well enough to plot our futures. We agreed we would start out in search of the White Army by way of Outer-Mongolia and Manchuria; then, when we found the Whites we would join our old aviation squadron.

With false identifications, our ragged clothes, and our invalid status, we were safe to travel by railroad to Novonikolayevsk, then south by rail

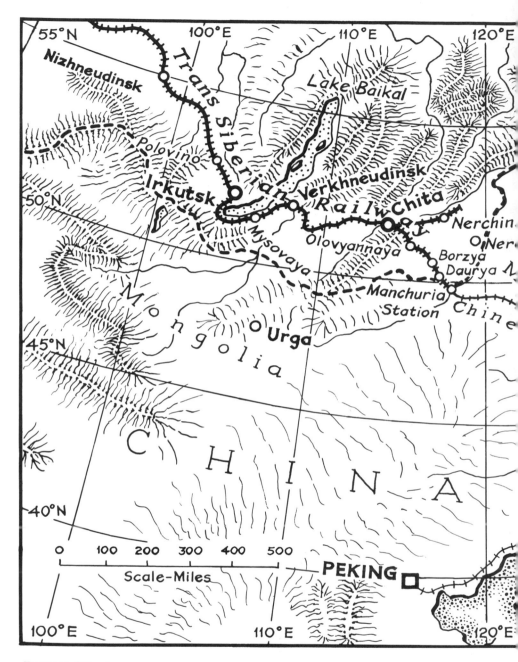

Eastern Siberia

This map illustrates the vast arena for the civil war between the Reds and Whites in 1920–1921. Alexander Riaboff, as a military pilot in various White squadrons, participated in this brutal struggle along the corridor of the Trans-Siberian Railway leading to Vladivostok. Riaboff's own journey culminated in his exile to Harbin, China, in 1920.

to Biysk. From there we proceeded leisurely south to the Altai Mountains until we found a village located on the northern slope of the mountains. We settled there for a while because we found a way to make a living by operating a forge that had been abandoned by a blacksmith. Muromtsev caught on to smithing very quickly and I became his assistant. All winter long we worked hard, accumulated food, clothing, and information about the mountain pass that would take us to Mongolia.

Early in April 1919, the weather was warm, and we decided to continue our journey. After dark one nearly-moonless night, we stole two horses, loaded our saddle-packs, and were on our way. We ran the horses at a goodly clip, so as to cover a safe distance from the village, until exhausted; then we stopped, tied up the horses, wrapped ourselves in blankets, and were soon sound asleep. I was awakened with a start by Muromtsev's cursing. "Hey, wake up—the horses—they're gone," he exclaimed in despair. Sure enough, we were stranded without means of transportation and we had a long, difficult distance to cover.

In a way, we were fortunate; the villagers could have killed us while we slept. The penalty for stealing horses is death, but they must have liked us well enough to have spared our lives, so long as they were able retrieve their stolen property.

At the base of the mountain, the spring weather had been warm, with new life bursting out all about us, but the higher we climbed, the colder it became, especially at night when the temperature dropped to zero or below. Snow fell one night and we knew then that we were in for a struggle just to survive. We walked as fast as possible to keep from freezing to death. Our frost-bitten, bleeding hands no longer had any sensation in them as we clawed our way over rocky crags. Being the stronger of the two, I had to prod, bully, and yell at Muromtsev to keep him on the move. To lie down and freeze to death would have been such a relief. We sustained ourselves on bread, cheese, dried fruit, and a bit of salt pork now and then. There was always snow to moisten our tongues.

At last, completely debilitated, we reached the summit. From there, we could see far into the Sinkiang plains and freedom! Gradually, on the south side of the mountains our bones thawed as the sun caressed our reviving bodies. Faster and faster we walked in order to reach the valley while the day was still young.

My friend, who had the keener vision of a pilot, suddenly grabbed my arm and said, "Look straight ahead into the far distance, there's a group of horsemen heading in our direction at full gallop!" Being unarmed, I should have removed my white undershirt, attached it to a stick and waved it, to show we were unable to defend ourselves. This, I am sure they would have understood, but it all happened so quickly and we had not expected to encounter hostility in Mongolian territory.

When the men were within shooting distance of us, they positioned themselves in a semicircle and began to fire their ancient muskets at us.

Only one man had a modern rifle, and he was the one upon whom I focused my attention. We could judge the timing of the musket shots and were able to dodge them, but when Muromtsev failed to look in the direction of the rifle bearer for just one second, he received a direct bullet to the heart. He slumped over, fatally wounded.

I felt as though I were being drawn straight to hell by some evil forces—my despair was indescribable. My dearest friend, my companion through trial and tribulation, my reason for struggling and wanting to survive, was leaving me forever. I know I cried out; beat the ground with my fists; tried to pray; pleaded with him not to die; then I wept, lying beside him inert, hoping they would shoot me too. Oh, how I wanted to die!

I felt a hand on my back and then a voice speaking in broken Russian, the way a Mongol speaks our language. "Who are you?" he asked. I told him, "We are White Russian officers escaping from the Reds. We want to join our aviation squadron in eastern Siberia." He looked at me pityingly, realizing the terrible mistake they had made. He and his men, he explained, had assumed we were Reds, whom they feared because of the violence they had encountered at their hands. "Shoot first, ask questions later" had become their motto.

They all tried to express sincere sympathy toward me, and their spokesman offered to assist me and make reparations for their error. I regained my poise very shortly; I had to do what was necessary—my friend had to be buried then and there; so we worked together to dig a grave deep enough to shelter the remains from predators. Part of me was interred that day too.

My newly found friends provided me with a horse, food, and clothing, and helped me to get started on my way across Mongolia. I crossed Mongolia, part of Manchuria and, at last, reached the Trans-Siberian Railroad in Harbin, and from Harbin to you. I could not have completed my journey without the generous help from the Mongols.

Unfortunately, Voshchilio's services as a navigator were not needed in our rapidly disintegrating aviation squadron. In a few days he left us for parts unknown and our paths never crossed again. Nevertheless, I have remembered his tragic story to this day—sixty years later.

Early in September 1920, we came to the conclusion that the civil war in eastern Siberia was coming to a close. A group of us met to discuss the matter of where to go. Seven of us officers who had wives, and a few who had children also, decided to go to Manchuria directly. The remainder chose to go to Vladivostok, Russia's far eastern port on the Pacific Ocean at the end of the Trans-Siberian Railroad, to offer their services in the continuation of the war.

Diary Excerpt, 1920:

The Retreat of the
White Army Continues

The fall of Omsk to the Reds in November 1919 required Alexander Riaboff to enter upon the most difficult period of his Siberian years. Most of this portion of his diary falls in 1920. These entries demonstrate that Riaboff was able to fly assorted missions even up to his last days in Siberia. He served for a time under the overall command of the brutal General Semyonov. There are also references to the Allied interventionists, American and Japanese, who played a significant role in shaping the course of the civil war in Siberia in 1920.

■　　■　　■　　■

December 25, 1919 *(at the city of Verkhneudinsk)*
My circumstances are so grim, I do not feel much like writing. Omsk fell to the Reds on November 15, 1919. With this defeat, General Sakarov replaced General Diderikhs as commander-in-chief, only to be removed in favor of General Kappel. There is much confusion. We hear rumors that Nikolayevsk and Tomsk also have fallen to the enemy. . . .

Now the news from the west is bad. General Yudenich, it appears, has been defeated outside Petrograd. Resistance to the Reds in the north has ceased. General Denikin is now falling back, surrendering Orel, Voronezh, Kharkov, and Kiev to the advancing enemy. Now the Reds are triumphant on all fronts. Even the Allies are refusing to continue their aid to us, leaving us to our fate.

The ruble is now worthless—an English pound equals twenty-five rubles, a pound of butter 150 rubles, and a quart of milk anywhere from 150 to 200 rubles!

There is even a political crisis within the territory still held by our side. General Semyonov acts independently at times, our parliament has refused any further aid to Kolchak, and the various urban cooperatives that have been organized are now falling under the influence of socialism. There is disorder everywhere.

I was ordered to accompany the housing officer of the Central Aeroplane Shops to Irkutsk, to find new quarters for our squadron. Once we arrived in Irkutsk, our Army Command suddenly decided to move my air unit to Verkhneudinsk. Upon arriving in the city, word reached us that yet another site for our squadron would be announced later. There is chaos and confusion everywhere.

The situation along the Trans-Siberian Railway has worsened. There is a critical shortage of coal, which on occasion has brought rail traffic to a standstill. One of our military supply trains, compelled to stop near the frontlines because of a shortage of coal, fell into the hands of the Reds. The railway station is now crowded with stalled railway cars. . . .

January 20, 1920 *(at the city of Chita)*
Our troop train has moved from Verkhneudinsk to Chita, another retreat eastward for our beleaguered army. When I reached Chita, I was attached temporarily to the Manchurian Aviation Unit of our army, under the command of Captain Pleshkov. We have four pilots and four observers in this unit.

This squadron will soon acquire French aircraft. This will be a welcome change because this small air unit lacks just about everything. We have few aircraft or adequate supplies. . . . This air unit rarely takes to the air!

The situation at the front is now approaching a crisis. The Reds continue to advance. No one seems to know exactly where the Reds are.

One week before Christmas there was a coup d'état in Irkutsk. The Social Revolutionaries* seized control of the city. They now advance two radical slogans: "Down with the civil war" (an attack on the Bolsheviks) and "War to the Counterrevolutionary East!" (against General Semyonov). It is clear to me that these Social Revolutionaries are, in fact, Bolsheviks. Their takeover of Irkutsk was aided by the Czech Legion.

Kolchak has appointed Semyonov as "Commanding Officer of the Trans-Baikal Region." Japanese interventionist troops have moved against the Social Revolutionaries, only to stop abruptly and declare their neutrality. The Allies have promised to solve this problem, but there is great confusion. Our troops, by some misunderstanding, have been disarmed in Verkhnyy Sinkhyan. The Czechs have moved through here on their way out of Russia. We hear that Kolchak . . . has been arrested in Irkutsk. Whatever remains of the Western Siberian Army, under the command of General Kappel, remains near Irkutsk. This army, it is hoped, will soon take Irkutsk again and join forces with General Semyonov.

All our aviation units were stationed at Irkutsk when the Social Revolutionaries staged their coup d'état. What happened then is unclear. The

---

*The Social Revolutionaries, led by Victor Chernov, were a radical revolutionary party known in the pre-1917 era for acts of terrorism. They advocated a form of peasant socialism.

Social Revolutionaries reassigned these air units to the west, toward the Polovino Station. No one knows where these squadrons are now, or what they are doing.

January 25, 1920 *(at Chita)*
Our French aircraft arrived here two days ago. These new airplanes are two-seaters and are powered by an eight-cylinder, 230-horsepower Salmson engine. This splendid airplane can climb to 2,000 meters in eight minutes. The armament consists of three machine guns. I am impressed with its speed and sturdy construction. My only complaint is that the aircraft is difficult to land. A total of twenty-three Salmson-powered aircraft have been delivered so far; our squadron obtained eight of them. There is a rumor that the First, Fifth, Seventh, and Eighth air units, along with a technical unit, will soon join us. Three new pilots arrived here, straight from flight school. As yet, we do not know who they are or what they stand for.

No news from Irkutsk. . . . Our army has retaken Verkhneudinsk from the Reds. The commander-in-chief of our Siberian forces is now General Semyonov. General Denikin has become the Joint Chief of Staff for all our forces. While the military situation remains unclear, things have stabilized. What a contrast to the tense situation of two–three weeks ago!

February 2, 1920 *(at Chita)*
Captain Astafiyev arrived here two weeks ago with a tragic story of how he escaped from Novonikolayevsk to Irkutsk. . . . The long trek to Irkutsk was arduous; he traveled by foot, on horseback and, on occasion, by train. Somehow, Astafiyev, accompanied by Lieutenant Dudka, reached Irkutsk. On the way, Dudka caught typhoid fever and was forced to remain in Irkutsk. Other members of my old squadron also reached Irkutsk in those tense days that followed (Dil, Voshchilio, Kompaniyetsev, and others). Captain Putsillo shot himself, Colonel Shapson, it is believed, was killed later in Irkutsk during the coup by the Social Revolutionaries. During this crisis, Makarov defected to the Bolsheviks.

Enormous quantities of war materiel were destroyed by us during this hasty retreat. The Reds captured other stockpiles of supplies. There were instances as well of simple theft in this chaotic situation. . . . The White Russian Political Administration is now located somewhere near Krasnoyarsk, but no one knows its exact location.

February 11, 1920 *(at Chita)*
There is a story circulating that General Kappel has died. Our new commanding officer is Voitsikhovskiy.

Political chaos is spreading. There are a number of cities that have fallen under the control of the Social Revolutionaries—Novonikolayevsk, Ussuriysk, Khabarovsk, Blagoveshchensk, and Vladivostok. Here in Chita everyone expects an uprising to come any day. There is a feeling of panic. Refugees from this political disorder are to be seen in many parts of the country.

The Japanese have arrived here in Chita. Their intentions are murky at best. Many wild rumors circulate. Some report that General Yevert, the Bolshevik leader who occupied Irkutsk, has now turned against his former commanders and is chasing Bolsheviks out of the Ural region. Some say Kolchak and General Popelyaiyev are dead. Others say they are alive. Reports continue to reach us about the Czech Legion, to the effect that twenty-seven trainloads of Czech troops have been destroyed.

This is perfect chaos!

Today our squadron tested a new American-made aircraft. Lieutenant Moshevitinov took the aircraft up for a test flight. The radiator boiled over, throwing steam which froze on his goggles. He managed to crash land the airplane (a total loss), and to emerge miraculously without a scratch!

We are now reassembling the new French aircraft. I.M. Dil has reached our air unit from the west.

February 22, 1920

General Vladimir Kappel has died, a victim of the Siberian winter. As a brave soldier, he froze to death in absolute disdain of danger. Yesterday his body was taken ceremoniously to the cathedral in Chita for the funeral. An enormous crowd witnessed the ceremony, consisting of the intelligentsia and several military units.

General Voitsikhovskiy arrived in Chita a few days ago as General Kappel's replacement. His army consists of around 50,000 men, now stationed in the Verkhneudinsk-Mysovaya region. Other White troops are arriving daily as we regroup.

Our army has been defeated not only on the battlefield, but by the never-ending series of retreats. These withdrawals have shattered our unity of command, broken communications, and prompted all kinds of disorder. At times our troops have been reduced to disorderly mobs. Many have perished under the constant attacks of the Reds.

Even before our army reached Krasnoyarsk, it had ceased to be a disciplined force. In fact, the army had been shattered into separate groups, each desperately trying to survive the rigors of a forced retreat across a vast and difficult terrain. At Krasnoyarsk, General Kappel did restore order and reconstitute a viable military organization.

Subsequent phases of our armies' retreat proved to be more planned.

When we approached Irkutsk and threatened to occupy it, the Czech Legion warned us to stay clear of the city. Rather than risk a confrontation with the Czechs, General Voitsikhovskiy sent his army around Irkutsk. This was a very difficult maneuver for his exhausted troops, a trek across frozen taiga forests through territory filled with Red partisans who made frequent and sudden attacks. Horses were the chief means of transport for all elements of Voitsikhovskiy's army. Almost no one escaped without frostbite.

The retreat of the White armies, which involved a march from the Volga River to Mysovaya Station, a total distance of nearly 3,500 versts, was epic. This march dwarfs Napoleon's retreat from Moscow. Eternal glory belongs to these brave soldiers! Their losses were enormous: when the army was at Novonikolayevsk, it numbered 75,000 troops; somewhere between 30,000–50,000 survived the long trek (this includes the garrison troops along the Trans-Siberian Railway that joined the retreat of the Whites eastward).

The long march, with its constant fighting and severe weather, made an impact on the Reds as well. They were in lock-step with the retreating Whites and suffered terribly, too. The Reds reached the point of disintegration even as we did. Once past Omsk, the Red Army—disorganized and decimated by typhoid fever—found itself faced with a severe crisis. Red troops began to drink heavily. Discipline broke down as Red soldiers started to live off the land, separating into small bands, to steal or rob, to do whatever necessary to survive. As a result, the Red Army has ceased to be a viable military organization. From Omsk all the way to Irkutsk, there is no effective government, only criminal bands—either Red or White—living off the land. . . .

Here in Chita, life ironically has taken on a serenity. Captain Astafiyev, Lieutenant Bocharev, and Lieutenant Starodubtsev have left my squadron. New arrivals include two pilots—Captain Pavlov, formerly the commander of the Fifth Air Unit, and Lieutenant Agapov.

February 29, 1920

Word has reached us that tomorrow, or the day after tomorrow, our squadron will move east. We are ready to move again. Our Manchurian Division, it is reported, will take a position in a region stretching from the Manchurian border to the Olovyannaya Station on the Trans-Siberian Railway. Chinsu and Verkhneudinsk will be occupied by General Voitsikhovskiy.

On February 26 our hangar burned down. Someone apparently tried to start the heating stove by pouring gasoline into it. Out of this unfortunate fire we salvaged some equipment.

March 19, 1920 *(at the Daurya Station)*

Our journey eastward began on March 1. After eleven days of travel from Chita, we reached the Maulskaya Station on the railway, about eighteen versts from Manchuria. We wanted to cross the border into China, but we were warned that Chinese troops would block any Russian train seeking to cross into Manchuria. As a consequence, we stayed at Maulskaya for five days, and then returned to Daurya. Here my aircraft was unloaded, reassembled, and prepared for flight.

Our situation has become desperate. The Chinese Railway, which provides a southern passage to Vladivostok, has been taken over by workers who are on strike. These workers openly support the Bolsheviks.

May 27, 1920 *(at Chita)*

Five months have passed. This interlude of silence seems like years.

At Daurya, I flew a French airplane, which proved to be an excellent aircraft except for its fast landing speed.

While at Daurya, I flew frequently. One mission called for me to fly over to Borzya, where I would receive orders for several combat missions. The flight from Daurya to Borzya was not a success. Once I took off, I forgot to turn off the pump that increases the flow of water in my engine. As a result, the engine overheated and I was forced to land near the Kharonor station. The landing was rough—I brought the aircraft down hard, breaking the wheel assembly and ramming the nose into the ground. Despite the violence of this uncontrolled landing, I had the aircraft repaired, reaching Borzya the next day. While flying out of Borzya, I made three long reconnaissance flights. . . . I then returned to Daurya.

After Easter, I received orders from our Aviation Command that I had been transferred to the Second Air Unit at Chita. I arrived back at Chita on April 23, had my aircraft reassembled, and was flying in a short time. My engine, however, continues to act up, despite constant attention and a thorough inspection.

Tomorrow our squadron will move up to the frontlines at Nerchinsk.

Things in Chita are quite different now. After news of Kolchak's arrest and execution reached Chita, the situation within the city became hopeless. The army units of General Kappel stood with elements of General Semyonov's army in a successful defense against a determined Bolshevik assault. Even the Japanese joined in the defense of Chita. The Bolsheviks now have retreated to the west. . . .

We have arrived, however, at the bottom of an abyss! Money no longer exists. There are severe shortages of food, military supplies, and even manpower to sustain the resistance. There is nothing! Money has lost its value—the Japanese yen goes for 1,300 rubles. . . . A heavy winter coat now costs 20,000 rubles. How can we go on fighting?

There is a rumor that the Japanese are now negotiating with the Bolsheviks.

June 10, 1920 *(at Chita)*

Since my return to Chita, there has been considerable combat flying for my squadron, now under the command of Captain Slyusarenko. We are flying almost every day and sometimes twice a day. We have had great successes with our reconnaissance flights, making contact with the enemy almost every time. . . . My assignment with the Manchurian Aviation Unit appears inactive by comparison. . . .

At Chita, military pilot Lieutenant Gusev developed engine trouble on takeoff, which forced him to crash-land near the airfield. Gusev's aircraft hit a wagon, killing two soldiers. Gusev died three days ago. Today they are going to bury him.

General Semyonov himself inspected our airfield on May 25. Just before the inspection a terrific windstorm passed over the field, damaging all the aircraft except the commanding officer's airplane. During an exhibition flight our commander took General Semyonov in his aircraft and I followed with Semyonov's personnel aide as my passenger. The inspection ended with a drinking bout in my tent.

The Bolsheviks are once again on the run. Our troops are mopping up the Reds in a sector between Sretensk and Karymskoye, near the Manchurian border.

It is apparent that the Japanese who are stationed around Chita are talking to the Reds. What they are negotiating is unclear, perhaps a scheme to build a buffer state that would extend westward to Irkutsk, or even to the Urals. There are many rumors.

Today there was an announcement, reporting an alleged coup d'état against the Reds in the Moscow region and Petrograd: "Trotsky has been killed and Lenin has fled to Novgorod. A new government headed by Patriarch Tikhon and General Brusilov has assumed power." I find this story to be quite unlikely!

The Japanese yen is now valued at 4000 rubles.

June 13, 1920

Two days ago I received an order to go on a liaison mission. The emergency order, given top priority, reached me at 7:00 p.m. One hour later I was airborne for a flight to Kavykachi-Gazimur against a strong headwind. After an hour and a half, I landed in complete darkness at the designated rendezvous point. Descending into darkness at this unfamiliar spot was a tension-filled maneuver.

Upon landing, I gave a thick envelope . . . to General Molchanov, who in turn gave me several thick envelopes to take back to my base. I returned the next morning. A Japanese officer, Major Yugami, flew back with me. By noon, I had reported to our staff on my flight. General Lokhvitskiy, in particular, took an interest in the materials from General Molchanov. Later, General Lokhvitskiy, after a long personal interview, promoted me to staff-captain. . . .

The Reds are falling back toward Nerchinskiy Zavod, where we plan a climactic battle against them on the fifteenth of this month.

July 31, 1920

Well, here I am in Borzya again! This time, however, the Japanese yen is worth 75,000 rubles!

The Japanese are evacuating the Trans-Baikal region. So are we. Our destination, however, is not certain.

We arrived in Borzya on July 15 after a trip from Nerchinsk for some rest. Over three days our aircraft were flown here from Chita. . . . While in Nerchinsk my squadron clocked 158 hours of flying time. I accounted for sixty hours, our commanding officer for about seventy hours, the rest divided between Agapov and Slyusarenko. Our superiors were very pleased with the jobs we have done at Nerchinsk. In fact, we were given a special dinner to honor our achievements, and a farewell dinner on July 10, just before our departure for Borzya.

A shameful event occurred one week ago. Lieutenant Kruchinskiy, Lieutenant Colonel Dydyudin, Ensign Agapov, and Warrant Officer Belyagin took four of our squadron's aircraft and defected to the Reds. Belyagin landed about thirty versts from Chita and then continued on foot. There are many other instances of defection to the Reds these days, on account of our hopeless military situation.

Tomorrow we may move on to Sharasut or maybe Harbin in Manchuria. Here in Borzya we have witnessed the near collapse of our army. One trainload after another of our troops have passed through here on their way east. But to where? There is news now that the Chinese are disarming our troops when they enter China.

We will join the retreat tomorrow. It seems everything will be over in a month or more. Our defeat has triggered disagreements among the top echelons of the White armies. . . . The news from the west is also grim. The Bolsheviks, we hear, are now at the outskirts of Warsaw. . . . General Wrangel has won some battles and is now going to the aid of the Poles and those peasants who have risen up against the Bolsheviks. . . .

Even our squadron is now gripped with dissension. The officers quarrel constantly. The soldiers do as they please. Colonel Kachurin, our present commander, has resigned in frustration. . . . If this continues, no one will remain.

August 13, 1920 *(at the Eighty-third Reassignment Area, near Daurya)*
Colonel Makarov, the chief of aviation, has refused to accept Kachurin's resignation. Kachurin continues to argue with Makarov on the running of the squadron. Soldiers obey no one, it seems, preferring to do what they feel like doing. Some soldiers have even talked about doing away with their officers. This kind of talk quickly ceased after a forceful intervention by Staff-Captain Beliaiyev. We are surrounded by heavy drinking, laziness, and a complete lack of discipline here in the reassignment area. Our camp is now divided into two groups, one satisfied with the present leadership, the other hostile and rebellious. . . .

Captain Bashkevich has left for Harbin with his mother. The older Slyusarenko, Colonel Andreiyev, and Warrant Officer Menziyelintsev are leaving the technical unit for unknown destinations.

Yesterday I flew back from Chita. It was my longest flight to date, a total of 400 versts. I went to Chita to pick up our passports, which will allow us to escape from Russia. Everyone, it seems, has obtained these passports. The evacuation of our troops continues. . . .

From now on we are being paid in gold ruble coins. For July and August I was paid around thirty-nine rubles. . . . One gold ruble is now worth 20,000 rubles in paper money. We are preparing for our trip out of Russia.

September 1, 1920 *(at the Eighty-third Reassignment Area, near Daurya)*
We are still here. Autumn has arrived early with high winds and rain. There is much despair in this camp, perhaps equalled only by our boredom.

My air unit has practically disintegrated. Rumors abound. There is much infighting among the officers, with all kinds of threats being exchanged. Colonel Bankovskiy, Colonel Shreider, and I were ordered by counterintelligence to report on these conditions.

I have traveled to Manchuria several times, to deliver messages to Astafiyev and Colonel Makarov. As a result, the Army Command has offered Colonel Kachurin the position of chief of aviation, and to me, the position of squadron commander. I accepted this promotion on condition that I would have freedom of action as far as the squadron was concerned.

Colonel Kachurin, however, has reneged on this promise. He has denied me command of the air unit. Always suspicious, Kachurin has launched an investigation of my air unit. This inquiry is based on reports we made to the Army General Staff on conditions in our squadron. The matter is being investigated by a Colonel Semyonov, who is a friend of

Kachurin. The officers in my air unit may be punished for expressing complaints.

I am now completely fed up with the military. Colonel Makarov, I understand, is issuing two-month furloughs—I intend to get one!

Colonel Kachurin was temporarily removed from his command. Kachurin got into trouble because four felons escaped from jail in Nerchinsk and stole the five horses belonging to our air unit. . . . Our rubles cease to exist as a means of exchange. Even in Chita they only accept gold, silver, or Japanese currency. Our aviation section is now severely depleted and is about to collapse as a functioning military organization. Lieutenant Pleshkov is intending to go on "leave," to escape this impossible situation. Colonel Kachurin also intends to abandon the unit, as I do. Only Lieutenant Colonel Mozzhevitinov, Ensign Demin, and Gros-Tonaker have stated they will remain.

September 10, 1920
Today I was ordered to fly to Sharasut, to pick up Captain Nagurskiy, and then to take him to Olovyannaya. I took off, but my aircraft developed engine problems. It was late in the day, so I returned to base. I tried again the next day, but short of Borzya, I had to return again because of an engine malfunction. I really didn't feel like flying anyway.

The next day, my wife and I left for Harbin.

9. Final Days

The final section of Alexander Riaboff's memoir deals with his passage into exile in 1920. The decision to go into exile reflected the impossible situation of the Whites, who by this date had been weakened mortally by Red Army offensives, disease, and starvation, and the disintegration that came from constant retreat across 6,000 miles.

■　　■　　■　　■

WE BOARDED A TRAIN EN ROUTE TO Harbin, Manchuria, on September 12, 1920. We had foreign passports, which assured us of our right to leave Russia. It was a miserable day, just as gloomy as our moods. Some wept openly, others sat brooding silently, and I, through a drizzling rain, surveyed the entire scene around me, including the countryside we were leaving: the flat, semidesert land with little vegetation, extending to a treeless horizon without a sign of human habitation anywhere. Yet, this was our native land, for which we felt such indefinable devotion. The thought of leaving the soil of Russia, never to return, filled me with utter despair. Holy Mother Russia! No longer did Russia bear that title, but I thought of her in that way. Now, as the train moved toward China, I realized that I was about to sever those ties that bound me to Russia and my family in Moscow.

Two days after crossing the border, we arrived in Harbin, with exactly sixty gold and silver rubles, the clothes on our backs, plus a small collection of personal possessions. The most turbulent, exciting period in my twenty-five years of existence had come to a close.

Very soon after our arrival, I was hired as a tutor by a well-to-do Russian family, who provided us with board and room as well. I felt fortunate to have landed a job without delay and, for the time being, I was quite content to put to use my thorough education in mathematics and physics.

When Sonya and I were about to become parents, I found a position as an appraiser for the Harbin Real Property Tax Department, which was

The Riaboff family in Harbin, China. This photograph was taken on June 23, 1923, just three months before the Riaboffs sailed for America.

more remunerative and permitted us to move into our own apartment in a comparatively new building with modern facilities.

Our daughter, Helen, was born on June 8, 1921. With this blessed event, we became a genuine family unit with new responsibilities, and it made me more alert to what was happening in Harbin.

The part of the Trans-Siberian Railroad that crossed Chinese territory was taken over by the Chinese government, and control of the city as well. Life immediately became more difficult for everyone. I also noticed that there was a steady increase in the numbers of representatives of Japanese commercial enterprises coming to Harbin. This indicated to me that the Japanese military would soon follow, under the pretext of protecting their nationals, but actually for the purpose of taking over Manchuria.

I said to Sonya, "My dear, we must leave Harbin and emigrate to another country. I predict that under Japanese domination, life for all of us will become intolerable."

"What country do you suggest?" she asked. "The United States of America, of course," I replied.

"Why, my goodness, it's so far from Russia," she exclaimed, starting to cry.

"What difference can that make? We'll never dare set foot in Russia again because of my defection from the Reds!" I reminded her.

"Why, of all countries, the U.S.?"

"That's simple to explain. In Germany, we would be forced to become Germans; in France, Frenchmen, but in the U.S.A., we'll be Americans, just like everyone else!" So, it was settled.

The three of us left Harbin for Japan, where we boarded the *President Jackson* and sailed for the United States. We arrived in Seattle, Washington, on September 1, 1923.

Diary Excerpt, 1920–1922:
Life in Exile

The Riaboff diary ends with a series of entries covering two years' of exile in Harbin. The desperate struggle to escape the triumphant Red Army in 1920 gave way to an equally difficult émigré life in China. Mirrored in Riaboff's diary entries are images of a large community of Russian political outcasts struggling to find their way in an alien society.

Riaboff provides here a candid account of his motives for leaving the White Army after two years of service. His references to his homeland suggest the profound attachment to Russia that he and his compatriots carried with them into exile.

■　　　■　　　■　　　■

September 1920 *(at the city of Harbin)*
We arrived in Harbin, China, this month with Colonel Shreider and his wife, Colonel Bankovskiy and his wife and their two children, Staff Captain Ovchinnikov, Ensign Chernik and his wife and child, along with Lieutenants Vasiliyev and Zadorin and their families.

Our trip across Manchuria from Russia was uneventful; nothing interfered with our travel or threatened us along the way. Now we are all beginning to adjust to our new lives in Harbin. Sonya and I have found a room and I am now looking for a job.

I am a free man in Harbin!

October 22, 1920 *(at Harbin)*
I have abandoned the armed struggle against the Bolsheviks. Yes, I am a deserter of sorts. Many things contributed to my decision to leave the disintegrating White armies—the absence of a concrete political ideal, the sense of impending defeat, and the gradual realization that continued struggle against the Bolsheviks would not end in triumph.

I resented the so-called "tactics" of General Semyonov, who "frittered away" precious time in remote localities. There were the disgusting living conditions, my meager salary, as well as countless other factors that made life extremely difficult. I guess it is better to become a deserter in this hopeless struggle than to continue as we were, living off the land, as "parasites," unable to do anything for the people.

May 31, 1921 *(at Harbin)*

More than half a year has lapsed since I last wrote in this diary. . . . Many exiled Russian officers and soldiers are now gathered in Harbin and other Manchurian cities. Some of these men have been able to establish a new life here and are doing well; others, less fortunate, are going hungry. Several mutual-aid associations have been formed to assist the most needy.

Russian refugees are now facing difficulties in many parts of the world—in Bulgaria, Turkey, Romania, and Serbia. Vestiges of the armies of Wrangel and Denikin have escaped to these countries, seeking new homes. Serbia, it appears, has been the most hospitable; Russians have been greeted as brothers there.

April 23, 1922

Almost another year has passed. Last fall and winter, there was a brutal famine in Russia. The Volga region, the Urals, parts of western Siberia, the Ukraine, and most regions in the bread-producing belt of the country have fallen victim to this deadly scourge. Many people have left their ancestral homes out of desperation. There is even cannibalism. Millions have died. Almost all of the aid has come from America, organized by [Herbert] Hoover. This crisis, alas, has deepened because the transportation system in Russia has been nearly destroyed; often it is impossible to deliver food, even if available, to stricken areas as a result of the destruction of the transportation system. This famine has occurred as a consequence of a natural drought, forceful requisitions by the Soviets, and insufficient food reserves. The worst time was the winter of 1921–1922. Now it is much easier. . . .

Life for us in Harbin is getting more difficult. The economy here is depressed. Unemployment is increasing. I am thinking of emigrating to America, where I could begin anew my formal education.

Epilogue

For Alexander Riaboff, the United States offered a safe haven for his family, and the prospect of a new life removed from the dangers and hardships of Revolutionary Russia. Riaboff worked briefly in Seattle, as a foreman at the Boeing aircraft plant, but eventually settled in the San Francisco area.

He brought to his new home a keen interest in aviation, but rejected flying as a career, largely because of his responsibilities as a husband and father. In America, he decided, there must be a different kind of work, one that offered intellectual challenge and the possibility of a secure life.

Riaboff found work in San Francisco as a draftsman for a patent attorney. He soon discovered that he had an aptitude for drafting and a growing interest in patent law. His diligence and hard work earned him the plaudits of his employers, who urged the young Russian émigré to enter law school, from which he graduated in 1930 at the top of his class. After passing the California bar, Riaboff entered a highly successful career as a lawyer.

Living in the San Francisco area brought Riaboff into contact with an established Russian-American community, which in the 1920s had been swelled by refugees from Soviet Russia. Riaboff attended Holy Trinity Russian Orthodox Cathedral in San Francisco and participated actively in the Russian community. In 1928, Riaboff was joined by his younger sister, Tatiana. She later married Sergei Hitoon, a soldier who had also served with the Whites in the Siberian struggle.

The emigration of Tatiana Riaboff to America brought a partial reunion of Alexander Riaboff with his family. The separation from his parents and other two sisters, Anna and Panya, was permanent. When Riaboff joined the Whites in 1918, he feared for reprisals against his family, still living in Moscow. Life in revolutionary Moscow under the Bolsheviks was harsh and, for a family with links to the counterrevolution, there were real dangers.

Riaboff later discovered that the Cheka (the Bolshevik secret police) had paid a visit to his parents in 1918, asking questions regarding the whereabouts of their son, an erstwhile pilot in the Red Air Fleet. But this gesture, so unwelcome and threatening in its implications, proved to be an isolated incident, not a prelude to arrest and imprisonment. No further harassment followed.

Alexander Riaboff, his wife, Sonya, and young daughter, Helen, in Manchurian exile.

Riaboff's parents, in fact, survived the rigors of the Bolshevik Revolution notwithstanding their "bourgeois" class identification and the fact that their son was serving in the White armies. When the workers took over the manufacturing plant where Vassiliy Riaboff, the father, had been sales manager, he was not harmed. He subsequently found work in the new Bolshevik government as an accountant. He died in 1924. His wife, Eudoxia, lived until 1946, suffering through the additional hardships of the Stalinist years and World War II. Both of Riaboff's sisters who remained in Soviet Russia endured these difficult years as well. Despite the geographical and political chasm that separated Alexander Riaboff from his family, there were opportunities for correspondence to maintain a tenuous link of communication.

For Alexander Riaboff, war and revolution had propelled him along unfamiliar paths, enduring extreme dangers and later, by necessity, embracing the grim reality of émigré life. One pleasant and unanticipated aspect of those early cruel years had been meeting Sonya Nikitina, who became his wife and shared the burdens of the long trek across Siberia into exile. Both emerged from the experience with an abiding opposition to communism, seeing in the horrors of the Stalinist period a continuation of the inhumanity they had witnessed in the years of the Revolution and the civil war. Yet, this opposition was tempered by a realization that Soviet rule, even with its excesses, had brought certain concrete improvements to their former homeland.

Despite profound feelings of opposition to the Communist political system, they remained deeply attached to Russia in all its difficulties. When Nazi Germany invaded the Soviet Union during World War II, the Riaboffs assisted in mobilizing medical supplies in America for shipment to besieged Soviet Russia.

Sonya Riaboff died in March 1945, just before the end of World War II. In December 1947, Alexander Riaboff married Alice Child Felix, a widow and the mother of three children. This second marriage brought a renewed sense of stability to Riaboff's life, and the birth of a second child, Peter Alexandrovich Riaboff. Both of Alexander Riaboff's children pursued professional careers: Helen Riaboff Whiteley, the daughter from his first marriage, entered the academic field, becoming a professor of microbiology at the University of Washington; Peter Riaboff became a dentist.

After four decades of exile, Riaboff visited the Soviet Union in 1960. Further trips followed in 1963, 1968, and 1976. His return to Moscow brought an emotional reunion with his two surviving sisters, Anna and Panya, and an opportunity to see his beloved homeland. He met his sisters and their families, visited the still-functioning and crowded Orthodox Church he had attended as a youth, and even located an old schoolmate still living in Moscow.

The decision by Alexander Riaboff to write his memoirs stemmed from his visits to the Soviet Union. The return to his native land served as a stimulus for him to see more clearly the tragic consequences that often accompany war, revolution, and civil war. He finished the draft for his memoir in 1980, after several years of thought and periods of writing. By completing the memoir, now combined with his diary and a number of his unique photographs, Riaboff sought to make a statement about human folly—the personal conclusion that no political cause can justify war and human suffering.

Alexander Riaboff died in February 1984.

Bibliography

Baedeker, K. *Russia, with Teheran, Port Arthur, and Peking.* Leipzig: Karl Baedeker, 1914.

Baerlein, Henry. *The March of the Seventy Thousand.* London: Leonard Parson, 1926.

Boyd, Alexander. *The Soviet Air Force Since 1918.* London: Macdonald, 1967.

Bradley, John. *Allied Interventions in Russia.* New York: Basic Books, 1968.

Chamberlin, William Henry. *The Russian Revolution 1917–1921* (rev. ed.). 2 vols. New York: Macmillan Co., 1957.

Chernov, Victor. *The Great Russian Revolution.* Translated and abridged by Philip E. Moseley. New Haven, Conn.: Yale University Press, 1936.

Daniels, R.V., *Red October: The Bolshevik Revolution of 1917.* New York, 1967.

Deniken, A.I. *The White Army.* Translated by Catherine Zvegintzov. London: Jonathan Cape, 1930.

Dwinger, Erich. *Between White and Red.* New York: Charles Scribner's Sons, 1932.

Fitzlyon, K., and Browning, T. *Before the Revolution: A View of Russia under the Last Tsar.* Woodstock, N.Y.: The Overlook Press, 1978.

Fleming, Peter. *The Fate of Admiral Kolchak.* London: Rupert Hart-Davis, 1963.

Florinsky, M.T., *Russia. A History and an Interpretation,* 2 vols. New York: Macmillan Co., 1953.

Footman, David. *Civil War in Russia.* Westport, Conn.: Greenwood Press, 1975.

Graves, William S. *America's Siberian Adventure 1918–1920.* New York: Jonathan Cape & Harrison Smith, 1931.

Hardesty, Von. *Red Phoenix: The Rise of Soviet Air Power, 1941–1945.* Washington, D.C.: Smithsonian Institution Press, 1982.

Hitoon, Sergei E. *The Noble Piglets.* Sacramento: Citadel Press, 1981. (Russian ed.: *Dvoryanskiye porosyata,* 1975)

Hodgson, John E. *With Denikin's Forces.* London: Lincoln Williams, Temple Bar Publishing, 1932.

Kennan, George F. *Russia Leaves the War.* Princeton, N.J.: Princeton University Press, 1956.

_____. *Russia and the West under Lenin and Stalin.* Boston: Little, Brown and Co., 1961.

Luckett, R. *The White Generals. An Account of the White Movement and the Russian Civil War.* London: Longman, 1971.

Moorehead, Alan. *The Russian Revolution.* New York: Harper and Row, 1958.

Pares, Bernard. *My Russian Memoirs,* London, 1931.

Pasternak, Boris. *Doctor Zhivago.* New York: Pantheon Books, 1958.

Riasanovsky, N.V. *A History of Russia.* New York: Oxford University Press, 1963.

Reed. John. *Ten Days That Shook the World.* New York, 1919. (Many reprints.)

Salisbury, H.E. *Black Night, White Snow: Russia's Revolutions, 1905–1919.* New York, 1978.

Schwartz, Harry. *Tsars, Mandarins, and Commissars: A History of Chinese-Russian Relations.* Philadelphia: J.B. Lippincott Co., 1964.

Stewart, George. *The White Armies of Russia.* New York: Macmillan Co., 1933.

Treadgold, Donald W. *The Great Siberian Migration.* Princeton, N.J.: Princeton University Press, 1957.

Trotsky, Leon. *History of Russian Revolution.* 3 vols. New York, 1932–33.

Ward, John. *With the Die-Hards in Siberia.* New York: Cassell and Co., 1920.

Wildman, A.K. *The End of the Russian Imperial Army: The Old Army and the Soldiers' Revolt, March–April 1917.* Princeton, N.J.: 1979.

Williams, H.W. *Russia of the Russians.* London: Pitman and Sons, 1914.